SEVEN
SKELETONS

ALSO BY LYDIA PYNE

Bookshelf

WITH STEPHEN J. PYNE

The Last Lost World

SEVEN SKELETONS

THE EVOLUTION OF THE WORLD'S MOST FAMOUS HUMAN FOSSILS

LYDIA PYNE

VIKING

VIKING
An imprint of Penguin Random House LLC
375 Hudson Street
New York, New York 10014
penguin.com

ISBN: 978-0-525-42985-2
978-0-698-40942-2 (e-book)

Printed in the United States of America
1 3 5 7 9 10 8 6 4 2

Set in Carre Noir Light
Designed by Nancy Resnick

TO STAN

CONTENTS

SEVEN
SKELETONS

FAMOUS FOSSILS,
HIDDEN HISTORIES

T he first time I ever met a celebrity was on a June winter morning in Johannesburg.

I was an undergraduate, a student at a paleoanthropology
field school in northern South Africa. As part of the summer's human
paleontology curriculum, we attended a lecture at the University of the
Witwatersrand given by the school's eminent scientist, Professor Philip
Tobias. For his talk, Professor Tobias had pulled out several well-known
fossil specimens from the university's fossil vault, setting them on flat
wooden trays atop red velvet, showing them off like rare gems awaiting
our appraisal as we filed into the room to take our seats. As students, we
all had seen casts of these fossils before, but these were The Real Thing.

Professor Tobias was a thin, petit man with carefully combed white
hair and a meticulously knotted tie. (At a modest five foot four, I felt as if
I towered over him.) He arrived for the lecture in a starched white laboratory coat clutching a small wooden box that he put at one end of the
lab bench. He began his lecture by describing several of South Africa's
well-known fossil hominins, or human ancestors—picking up one of the
hominin specimens in front of him, turning it over in his hands, pointing

out an anatomical feature on the bone, and then carefully putting each fossil back on its tray. The man exuded gravitas and scientific solemnity. The fossils we were looking at represented decades of research and epitomized the crucial role that South Africa plays in understanding human evolution. As the stories about the different fossils seamlessly flowed together, it was obvious that Professor Tobias had given this lecture many times before, but *we* had never heard it. We were entranced.

But the specimen everyone was particularly keen to see was the Taung Child—a fossil whose history has loomed rather larger than life in the science of paleoanthropology. Ever since its discovery in 1924, the story of Taung Child has been chock-full of heroes, villains, theories, petty feuds, and the search for "scientific truth." Historical tradition champions the perseverance of the fossil's discoverer, Dr. Raymond Dart, in his belief that the fossil was, in fact, a human ancestor and not some aberrant type of fossil ape—an argument that ran against the grain of the scientific establishment in the early twentieth century. When Dart's beliefs were eventually accepted by the scientific community, the story of his dogged belief in his fossil practically became catechism in paleoanthropology that good science will be ultimately vindicated in the face of skepticism.

Back at the fossil demo, Professor Tobias eventually worked his way down to the old wooden box at the end of the table and, with a twinkle in his eye, pulled it closer. Drawing out the anticipation, he finally opened the box with theatrical flair. Reverently, he pulled out a tiny cranium and jaw. The pieces were small, gracile, and easily fit into Tobias's weathered hands; he told us that the wooden box was the same one that Raymond Dart himself had used to store the fossil at the University of the Witwatersrand for decades. After recounting the story of how Dart, who'd been Tobias's own academic adviser, found the fossil in a box of breccia from the Buxton Limeworks mine, Tobias put the fossil pieces together so that the lower jaw rested under the Taung Child's tiny face.

The fossil stared out at us, sizing up our group. Professor Tobias moved the little mandible up and down, clicking the fossil's tiny front teeth together, and launched into a well-rehearsed comedy act of sorts that had the Taung Child telling a few jokes, commenting on the weather, and offering some insights about the early days of paleoanthropology with his good pal Raymond Dart. This ventriloquy was met with stunned, shocked silence.

The veneration that had surrounded the fossil only moments before when Tobias described its historical significance now seemed oddly out of place. To us earnest undergraduates, it was like vaudeville. How could someone as respected as Professor Tobias show off something as famous as the Taung fossil that way?!? This wasn't the way we were *supposed* to experience it! The fossil should be in a vault. Or a museum display. Behind glass. *Anywhere* but auditioning as the straight guy for a Laurel and Hardy act.

Professor Philip Tobias holding the Taung Child. Fossil lecture, University of the Witwatersrand. *(L. Pyne)*

Over the last century, the search for human ancestors has spanned four continents and resulted in the discovery of hundreds of fossils. While most of these hominin finds live quietly in museum collections for experts to study, there are a few fossil ancestors, like the Taung Child, and others, like Lucy, that have become world-renowned personae, celebrities in their own right. These fossils live lives apart from their museum shelf and catalog number—because they are ambassadors of science that speak to nonexpert audiences, they have been given enough cultural cachet to transcend their status as scientific discoveries. Although the methods of scientific inquiry have changed significantly in the last hundred or so years of paleoanthropological research—to say nothing about the fluctuations in research questions and scientific paradigms—these celebrity fossils still remain part of a cultural gestalt. The fame and importance of these fossil ancestors mean that they are more than just the sum of their science; they play an important role in how audiences interact with scientific discoveries.

But what makes one discovery a celebrity and not another? Why does one fossil hominin come to have a nickname, museum exhibits, or even a Twitter handle while others simply sit in a museum drawer? And why might the answers to those questions depend heavily on the fossil's own cultural narrative? "Skulls or remains can tell only part of a story. Bones are mute," anthropologist Kopano Ratele suggests. "Stories must be told about them. They must be talked about, lectured on, explained, worshipped, searched for, retrieved, commemorated, archived, drawn, photographed and represented to restore their meaning. Knowledge must be built around them."[1]

In order to understand a fossil's celebrity, it's important to understand what the fossil is, where it comes from, and what kind of context it lives in. In other words, we need to situate the fossil within its own cultural history, giving it a biography built out of museums, archives, media, people— the countless interactions throughout the fossil's life after its discovery.

———

Every fossil story is about life and death. Fossils form when plants and animals die and their remains—bones, in the case of animals—are preserved in their surrounding geologic circumstances, a process that takes

Cast of the Taung Child fossil displayed in hands-on exhibit at the Origins Centre, University of the Witwatersrand, 2013. *(L. Pyne)*

thousands, sometimes millions, of years. Not all settings preserve fossils equally well. Some geological settings and landscapes are much better than others for preservation, and these areas are prized by scientists because excavations have better chances of yielding fossil discoveries. Not only are certain types of rock better than others for fossil preservation—limestone, for example, is a sedimentary rock that preserves specimens particularly well—certain landscape settings are more conducive than others for protecting the context of fossils after the organism dies. The successful interpretation of fossils hinges on understanding the kind of rock and landscape that surrounds them—being able to find fossils and then being able to appropriately contextualize them. Fossil hominins—those extinct species, ancestral taxa, with a close evolutionary relationship to modern *Homo sapiens*—can be particularly tricky to find and even complicated to make sense of.

The business of finding fossils—particularly fossil hominins—has a long and complicated history. Some fossils are found through almost haphazard circumstances and others through meticulous excavations. The very first fossil hominin specimens were discovered in the nineteenth century, although few of these early discoveries were recovered through methodical excavations. The hunt for human ancestors really began during the early twentieth century when the scientific community's hominin discoveries were popularized through newspapers, museum exhibits, and the occasional parody. Even today, the discovery of hominin fossils isn't a given thing, and the nature of fossil discoveries is highly variable. Many fossils are found by field survey or happenstance, where other discoveries are the result of decades of systematic research at one particular site in one particular region, concentrating on fossil locales that fit a researcher's specific agenda. Moreover, scientists can work at a site or in a particular area for a long time before anything is discovered, even when working at sites that have produced fossils in the past.

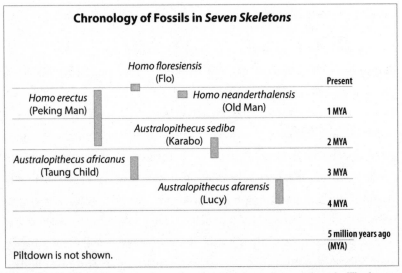

Illustration of fossil hominin chronology, with time scale on right side. The longer the box, the longer the fossil species appears in the geological record; each of the famous fossils is listed with its corresponding species. Since Piltdown is not a real fossil, its species does not have a correlating geological time span. *(L. Pyne)*

But the discovery of a fossil hominin is only the first step in understanding a fossil's evolutionary significance. From its field discovery, the fossil will generally go to a laboratory or museum associated with the research project for cleaning and there it will be assigned a catalog number, making the fossil part of a museum collection. Scientists will study the fossil and compare it with other similar finds. Measurements and photographs can be taken and analyses conducted. Although some of the technologies for comparing fossils have changed since the early twentieth century (we've traded stereoscope slides for CT scans), comparison is still the foundation for characterizing new fossil discoveries. From Neanderthal remains discovered in the late nineteenth century to the latest fossil Denisovans recovered in the twenty-first, all fossils must be described and put into their respective contexts. At this stage, context means a description of the fossil's geology—the type of sediment and rock that surrounded the fossil—as well as any archaeological artifacts, like stone tools, beads, or pigments, associated with the bones.

These initial studies are reported in scientific journals, and from there the life of a fossil can take a plethora of different paths. Some will simply go back to drawers and shelves in a museum laboratory; these specimens will continue to be studied and offer valuable insights into future scientific studies, but will only be described as part of a table of data—never as something singular or unique. Other fossils will have casts created for museums and laboratories so that it is easier for other scientists to see the specimen without having to ship the original to different locations. Some fossils will be brought into the media spotlight. For some particularly exciting finds, discoverers might hold press conferences to introduce their fossils to the public. Reconstructions of what the hominin might have looked like might head to museum exhibits. Scientific studies will continue. But the afterlife of these fossils isn't predetermined and depends on a number of factors. This is where some fossil discoveries will go on to become cultural touchstones, illuminating important moments in the history of paleoanthropology, and for other fossils, that celebrity will never happen.

———

Thinking back on my undergraduate encounter with the Taung Child, I would venture to guess there are many, many students, scientists, researchers, and visitors who saw Dr. Tobias show off the Taung fossil with his same comedic shtick. While Dr. Tobias was alive, watching him pull the Taung Child out of its box and clack its teeth together was— and telling the story of this performance still is!—just as much a part of the fossil's life as reading about its discovery and scientific controversy. This particular experience is just as integral to the Taung Child's identity and history as its scientific papers and museum publications.

It is easy to think that a fossil's importance comes strictly from its scientific value. Scientific significance is certainly *a* reason for fame, but it is not *the* only reason. Some fossils are famous for being "the first," "the most," or "the earliest" of something. Some are famous for the mystery and intrigue that surround them. Some are icons. Some are fake. Some are forgotten. Some fossils are famous, to echo cultural critic Daniel Boorstin, simply for being famous. All famous fossils, however, are fundamentally shaped by their various audiences, and as the audiences and contexts change, so does the nature of the fossils' celebrity. All celebrity fossils have a flashpoint where science, culture, and history intersect to ignite their fame. The fortunes of these famous fossils live and die within their cultural provenience—within their contexts and their histories.

This is particularly true in how we personify fossils, especially fossil hominins. To be a successful celebrity, the specimen moves from "only" being a famous object (an "it") and moves into becoming a "him" or a "her." It gets a moniker—a nickname—and a persona that become the cultural shorthand for the historical, physical, and psychological factors within it. Through the simple bestowing of a name and a pronoun, we are in effect granting the fossil agency, likability, and even a moral dimension. "Celebrity is made by simple familiarity, induced and re-enforced by public means," argues Boorstin. "The celebrity therefore is the perfect embodiment of tautology; the most familiar is the most familiar."[2] We judge the fossil by the stories we tell about it, and famous fossils are stories of heroism, notoriety, and celebrity. Since fossils lack any sort of intrinsic agency, their significance comes from the people and cultures that surround them. We shape their stories of fame today, just as historical forces shaped their interpretations in the past. When we understand the stories of these fossils, we see how science, history, and popular culture interact to produce celebrity scientific discoveries— that intersection means that these fossil human ancestors become cultural yardsticks through a huge number of material texts.

These bits and pieces of material life surround each of the seven famous hominin fossils in this book. These are fossils with postcards, formal portraits, curated exhibits, T-shirts, and posters. (Gentle reader, I have even seen a set of nail clippers in a visitors' center gift shop with the likeness of Mrs. Ples, well-known fossil from South Africa, enameled on the top.) The ephemera that surrounds a celebrity fossil is a part of its social archive and part of its own cultural identity.

———

But this circles back to the question of why certain fossils become famous. Which ones can and do achieve status as superstars? And what kind of celebrity cultural history would set one fossil apart from others?

"How could you write a book about famous fossils and *not* write about Mrs. Ples?!?" a horrified colleague asked me when I outlined this book idea to her, listing the fossils I planned to write about. "Or Ardi? What about Java Man, from 1891? Or, wow, *any* of the fossils that the Leakey family has found in decades of working in East Africa?!? How could you *not* include those???" She was polite enough not to follow her line of questioning with, "What kind of book is this?!?"

The question is certainly a fair one. What makes these seven fossils that I profile here famous in ways that differ from the many, many other specimens that fill laboratories, collections, and museums? These other fossils carry scientific significance and cultural import . . . why haven't they reached the celebrity echelons that these seven enjoy?

I've chosen to write the biographies of seven fossils that I feel tell us how scientific discoveries become written into popular culture and scientific ethos. These fossils are born through fantastical stories of discovery and live their lives, successfully resonating with audiences over decades. "[Museum specimens'] fame in life and their iconic status in death defy taxonomy," notes museum historian Samuel Alberti. "They are not only specimens, but also personalities; not only data, but also historical documents."[3] In other words, the stories and traditions associated with these fossils—their cultural identities—can't be separated from the story of who interprets them and how they acquire meaning.

These types of famous fossils are given pithy nicknames, written into evolutionary story lines, aggressively marketed, and, arguably, easily become cultural touchstones. As these fossils pop up in everyday media, as they live in museum exhibits, and as they continue to provoke deep scientific questions, their bones make cultural demands. Kopano Ratele suggests: "To become part of a culture, discipline, or project, bones need interpreters—paleontologists, painters, sculptors, kin."[4] These famous fossils help us understand how we make sense of our fossil ancestry.

Not only are these seven fossils all celebrated discoveries, each fossil illustrates a different kind of fame or notoriety within scientific and public circles. Lucy has become an icon; the Taung Child a folk hero. The Old Man of La Chapelle established the cultural archetype for Neanderthals. The Piltdown Hoax became a cautionary tale about preconceptions in science. The Peking Man fossils from Zhoukoudian took on a dramatic flair of paleo-noir as the fossils were lost and have never been recovered, vanished into legend like the Maltese Falcon. Flo is inexorably linked with her hobbit-infused identity. And the most recent fossil celebrity, Sediba, has embarked on a public relations campaign to become a fossil of serious scientific repute since its publication in 2010. These fossils are vivid examples of how discoveries have been received, remembered, and immortalized and serve as reminders of how our past as a species continues to impact, in astounding ways, our present culture and imagination.

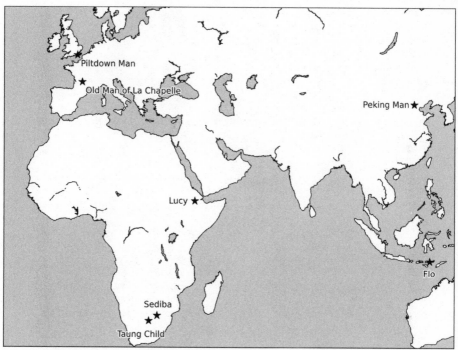

Map of the seven fossil discoveries. *(S. Seibert)*

These fossils live rich, vibrant lives, even though they are nominally tucked into their vaults at various museums. These seven tell us about our evolutionary ancestry—they give us millions of years worth of details about adaptations, selection pressures, even paleoenvironments—that preceded *Homo sapiens*. They demonstrate that science is a social and cultural process—how hypotheses are evaluated, how theories change, how technology is an ever changing tool for creating knowledge. As the stories of these fossils are told and retold, adding layer upon layer of cultural meaning, their histories become ever more entwined with our own.

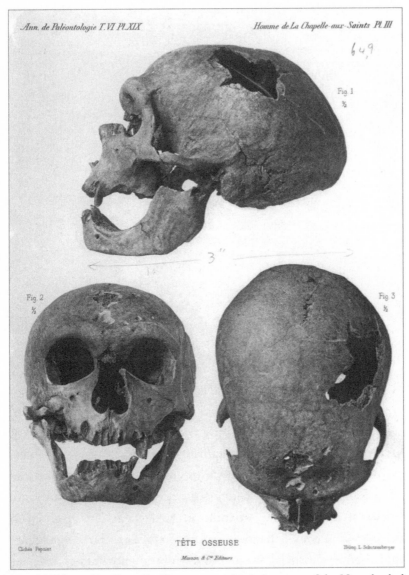

649

Fig. 1
½

3"

Fig. 2
½

Fig 3
½

TÊTE OSSEUSE

Clichés Papoint   Héliog. L. Schutzenberger

Masson & Cⁱᵉ Éditeurs

The Old Man of La Chapelle. These pen-and-ink drawings of the Neanderthal were created by Monsieur J. Papoint, under the direction of Marcellin Boule, and printed in Boule's *L'Homme Fossile de La Chapelle-aux-Saints,* 1911.

CHAPTER ONE

THE OLD MAN OF LA CHAPELLE:
THE PATRIARCH OF PALEO

O n August 3, 1908, a curious skeleton was discovered in south-central France by three French abbés. These abbés, Amédée Bouyssonie, his brother Jean Bouyssonie, and their colleague Louis Bardon, were all experts in prehistoric archaeology and were conducting an archaeological survey of the caves around the small French village of La Chapelle-aux-Saints. Their survey would map and document new Stone Age sites—any artifacts recovered during the excavations had the potential to shed light on early human prehistory.

The Bouyssonies and Bardon began their work in July 1908 and the first cave they surveyed yielded stone tools and fossilized animal bones—artifacts that strongly suggested that the region was a perfect fit for their Paleolithic research agenda. Encouraged by this early success, the prehistorians redoubled their efforts and began excavations in a second *bouffia*, or cave. In addition to more artifacts and fossils, the archaeologists found something unprecedented in early twentieth-century Paleolithic research: they found a burial pit containing an intact humanlike skeleton. As workers peeled back the layers of dirt around the skeleton, the abbés saw that the body was flexed in a fetal position, knees pulled up to the chest.

Subsequent study of the skeleton in the ensuing years showed that the bones belonged to a man—a toothless old man who'd suffered from osteoarthritis. But the skeleton wasn't the remains of a very old *Homo sapiens*. The skeleton was that of a Neanderthal, an extinct species of near-human first discovered in 1856. Although isolated Neanderthal fossils had been turning up for decades in sites across Europe and North Africa, prior to the abbés' discovery in 1908, no complete Neanderthal skeleton had ever been recovered. Their discovery was quickly nicknamed the Old Man of La Chapelle, and the fossil has shaped, directed, and influenced scientific research as well as public perceptions of Neanderthals for over one hundred years.

Fifty years before the discovery of the Old Man, research in paleoanthropology and archaeological prehistory looked very different than it did in 1908, at the beginning of the twentieth century. Both paleoanthropology and prehistoric archaeology were rather newly forming scientific disciplines, focused on humankind's long evolutionary history. In the mid-nineteenth century, fossil discoveries were interpreted within a framework of natural history, drawing from natural history's methodology and theoretical framework. Consequently, the mid-nineteenth century was an exciting time for those interested in studying fossils and understanding species change and even extinction. For example, French naturalist Jean-Baptiste Lamarck argued for the inheritance of an organism's acquired characteristics. Charles Lyell popularized Scottish geologist James Hutton's theory of uniformitarianism in his own 1830–1833 publication of *Principles of Geology*. In 1859, Charles Darwin published *On the Origin of Species by Means of Natural Selection, or the Preservation of Favoured Races in the Struggle for Life.*

Botanical, biological, and ethnographic collections were gathered by explorers and amateur naturalists, demonstrating the wide range of life on earth. Museums proliferated, taking the cabinets of curiosities from older generations and creating formal institutions, giving these newly amassed collections of animals, plants, and fossils new life.[1]

By the end of the nineteenth century, natural history was a more structured intellectual endeavor than it had been for centuries, and all of these new scientific endeavors needed practical grounding from field research. Lyell published geological profiles that showed evidence of glacial movements in the Alps. Darwin bred pigeons as an experiment to gather evidence for his theory of evolution by natural selection. (He later bequeathed all 120 of his pigeon specimens to the Natural History Museum in London in 1867.) The mid-nineteenth century also marked the origin of Neanderthal research, beginning with the discovery of the species in the 1850s. Natural historians interested in the deep time of human past began looking to records of material culture—stone tools and artifacts—in systematic ways; it was the beginnings of what the twentieth century would call methodology for archaeology and paleo-anthropology. The stone tools and other artifacts gave substance and data to what new scientific theories—and new disciplines—meant for the *longue durée* of humanity's history.

The Neanderthal story begins in August 1856, when limestone quarry workers blasted out the entrance to the Feldhofer grotto in the Neander Valley of west-central Germany. Picking through the rubble, workers found a set of skeletal remains and turned the skull scraps, arm bones, ribs, and part of a pelvis over to one Johann Carl Fuhlrott, an amateur naturalist and local *gymnasiat* schoolteacher in Elberfeld. (The workers assumed the bones were from an ancient cave bear.) Fuhlrott's degree in natural sciences from the University of Bonn helped him to appreciate the uniqueness of the materials the workers had given him. Although he quickly recognized that the bones were humanesque (and not ursid), he noted that the bones were unusual.

The cranium was extremely thick and differed greatly in shape from that of a human skull. Moreover, the skull case was elongated and the brow ridges above the eyes were almost ridiculously pronounced. Fuhlrott thought that the bones were most likely very old, since they had high mineral deposition on them and their stratigraphic provenience—the location of their discovery in the cave's sediments—showed that they were not recent additions to the cave.

Fuhlrott's cursory examination of the bones led him to conclude that he wanted a second, more expert opinion, so he delivered the Neander Valley skeletal remains to famed professor of anatomy Hermann Schaaffhausen at the University of Bonn. Schaaffhausen was impressed by what he called the "primitive" form of the skull and the evidence for its geological antiquity. (Fuhlrott had carefully questioned the quarry workers in order to substantiate his claim that the geological context of the skeletal materials really was old.) According to both Schaaffhausen and Fuhlrott, the bones were legitimately old and definitely humanlike, but still very different from other skeletal remains of *Homo sapiens*.

In addition to his expertise in human anatomy, Schaaffhausen had the scientific connections to be able to introduce this curious find to a broader natural history community. Fuhlrott and Schaaffhausen publicly announced the discovery and description of the bones in June 1857 at a Niederrheinische Gesellschaft für Natur- und Heilkunde meeting, convinced that the Lower Rhine Medical and Natural History Society chapter, which met in Bonn, was the perfect opportunity to introduce their studies of the bones to an invested audience. Together, they argued that the bones represented an ancient race of humans that had belonged to the German area. "The human bones from the Neanderthal," the two wrote, referencing the region of the fossil's discovery, "exceed all the rest in those peculiarities of conformation which lead to the conclusion of their belonging to a barbarous and savage race."[2]

Indeed, Schaaffhausen argued in his presentation to the naturalist group that "sufficient grounds exist for the assumption that man coexisted with the animals found in the *diluvium* [biblical flood]; and many barbarous races may, before all historical time, have disappeared, together with animals of the ancient world, whilst the races whose organization improved have continued the genus."[3] Schaaffhausen argued that the bones belonged to an extinct race of humans, but he did not specifically argue that they belonged to a separate and distinct fossil species. In recent decades, paleoanthropologist Dr. Ian Tattersall noted, "In hindsight, one can see how tantalizingly close Schaaffhausen came to an evolutionary perspective on his fossils, for he cleverly worked his notion of the mutability of species into his argument."[4] Schaaffhausen published a paper on the Neanderthal fossils in the *Archiv für Anatomie, Physiologie und wissenschaftliche Medicin* in 1858, and in 1859 Fuhlrott published a paper in the *Verhandlungen des Naturhistorischen Vereins der preussischen Rheinlande und Westphalens* that described the geology of the Neander Valley site and recounted the story of how the bones were discovered. Both men believed the Neanderthal fossils dated from a period when extinct animals such as mammoths and the woolly rhinoceros still lived in Europe, which would make the fossils among the oldest human remains known.

It goes without saying that these fossils generated considerable debate in Germany and abroad. The prominent German anthropologist Rudolf Virchow rejected outright Schaaffhausen's interpretation of the fossils. Virchow considered the Neander specimen to be a pathological anomaly of a recently deceased human—he thought the anatomical oddities, such as the skull shape and brown ridges, could be explained without invoking the rhetoric of species mutability. Virchow, an antievolutionist who abhorred the idea of species change, was also a dominant figure in German life sciences at the time, so his criticisms carried weight. In addition to Virchow's skepticism, August Mayer, a colleague of Schaaffhausen's at the University of Bonn, constructed an even more peculiarly specific account of the Neander specimen's life history. Mayer argued that the bones belonged to a person who had suffered from rickets and whose constant frowning from pain formed the bony ridges above the eyes. Mayer suggested that Fuhlrott and Schaaffhausen simply had found the remains of a deserter from the Cossack cavalry who had stopped in the Rhine in 1814.[5]

The Neander Valley skeletal remains gained a more secure footing within the scientific community in 1863 when professor of geology William King of Queen's College in Galway, Ireland, presented a paper at the 1863 annual meeting of the British Association, known today as the British Science Association. King argued the Neanderthal fossils belonged to an extinct species of early humans, and King went one step further and declared the fossils to represent a new species, *Homo neanderthalensis*, a species that was distinctly separate from our own, *Homo sapiens*. (His talk appeared in print the following year.) Even eminent natural historian Thomas Henry Huxley championed the Fuhlrott and Schaaffhausen skull as a member of the Neanderthal species and noted that the skull was "the most pithecoid of known human skulls."[6] ("Pithecoid" here means "apelike.") Huxley estimated that the cranial piece was of "normal" capacity—in keeping with what might be expected in a human population—and suggested that the Feldhofer skull was much more similar to an Australian Aborigine than to any living ape population. With so much interest swirling around the bones, the Neanderthal species was gaining force and presence in natural history circles because it inspired so many kinds of research questions.

After the specimen from the Neander Valley—designated as Neanderthal 1, the type specimen of the species—created taxonomic validity for Neanderthals, museums across Europe began to reexamine their collections. Several specimens that had previously been declared oddities or aberrations of *Homo sapiens* were now folded into this new species and were designated as *Homo neanderthalensis*. As a new "almost-human" fossil species, Neanderthals provided the budding discipline of paleoanthropology with a plethora of fantastic specimens for study. There was a child's cranium from Engis, Belgium (discovered in 1829–1830), a female cranium from Forbes' Quarry, Gibraltar (originally discovered in 1848), as well as other skeletal fragments scattered throughout museum collections in Europe. Archaeological excavations to recover more such fossils began in earnest across Europe—particularly in southern France during the first decade of the twentieth century—and these new sites yielded a plethora of Neanderthal fossils. By the time that American Henry Fairfield Osborn, the paleontologist and director of the American Museum of Natural History, embarked on his grand tour of Europe in 1909 to see Europe's plethora of Paleolithic archaeological sites, dozens of Neanderthal specimens had been comfortably situated in scientific literature.

———

Once Neanderthals had been firmly established as a fossil species, the next challenge was to try to make sense of their place in an evolutionary schema. Where had Neanderthals come from? What had their culture and technologies looked like? And why did they die out? These questions carried an implicit juxtaposition of Neanderthals with modern humans. The questions also contained an implied claim about human evolutionary history: one species, ours, was successful—we survived to the present day—and the other was not. For early-twentieth-century researchers, this meant that there was something about humans—technology, culture, aptitude, some kind of something—that destined humans to "succeed" where Neanderthals had "failed."

In the decades after Schaaffhausen and Fuhlrott, King, and even naturalists Huxley and Darwin, a significant amount of effort went into evaluating the question of how human or not Neanderthals were. The interest in Neanderthals quickly spread across Europe, drawing out researchers from geology, paleontology, and natural history, to say nothing of prehistory itself. In France, researchers interested in the "antiquity of man" focused on exploring and excavating caves with good potential for Paleolithic materials to tease apart and piece together the narrative of prehistory.

By 1908, this interest in understanding more about Neanderthals—and in excavating Neanderthal archaeological sites—led the abbés Amédée Bouyssonie, Jean Bouyssonie, and Louis Bardon to the sites in the Dordogne region of south-central France, to caves near the small village of La Chapelle-aux-Saints. As prominent prehistorians, the Bouyssonie brothers were familiar with the archaeology of the region, and their survey would explore and excavate the region's extensive complex of caves, or *bouffias* as the features were locally called. Photographs of their survey show caves carved out of the region's gray-white limestone with plants growing in the mottled gray rock, draping themselves over the entrances to the caves. The slopes of the hills are scattered with scrubby foliage—few, if any, trails crisscross the scree slopes, but caves clearly dotted the landscape.[7]

That July, the Bouyssonie brothers, along with Bardon, excavated stone tool artifacts and small pieces of rhinoceros horn and vertebral fragments in the first cave they examined in the La Chapelle area. Encouraged by these initial successes, the three prehistorians turned their attention to a second *bouffia*. This cave had an unusual geologic mudstone feature—a marl—that ran close to the cave's entrance, suggesting that the cave was old enough to fit their archaeological interests. Initial excavations of this marled cave yielded bone and stone fragments, similar to what they had already uncovered in their first site. However, on August 3, the excavators began to uncover something even more exciting. They peeled back the cave's sediments to find a humanlike skull. The Bouyssonies and Bardon continued excavations and found the rest of the male skeleton, curled up in a fetal position. Had he died in the cave like that, his body eventually covered over time by the cave's sediments? Or had he been deliberately buried there? The abbés felt that what they found was, in fact, a burial. In

L'HOMME FOSSILE DE LA CHAPELLE-AUX-SAINTS.

Fig. 2. — La colline où s'ouvre la grotte de La Chapelle-aux-Saints (Phot. de M. Papoint).

Fig. 3. — Entrée de la grotte de La Chapelle-aux-Saints (Phot. de M. Papoint).

— 117 —

ANNALES DE PALÉONTOLOGIE, t. VI, 1911.

Photographs of La Chapelle cave prior to excavations, 1908. Printed in Marcellin Boule's *L'Homme Fossile de La Chapelle-aux-Saints*, 1911.

publishing the results of their excavation, Jean Bouyssonie described the body's context as, *"La fosse n'a pas une origine naturelle"*—The pit does not have a natural origin.[8] This unnatural origin, then, meant that the pit had been deliberately dug and the body purposefully placed in it.

Fig. 9. — Intérieur de la grotte de La Chapelle-aux-Saints. On distingue nettement la fosse où gisait le squelette humain (Phot. de M. Papoint).

Photograph of La Chapelle cave's archaeological excavations, 1908, with picnic basket for scale. Printed in Marcellin Boule's *L'Homme Fossile de La Chapelle-aux-Saints*, 1911.

Concerned about looters and trespassers, the archaeologists quickly finished their work excavating the skeleton in the La Chapelle cave. They loaded up the skeletal remains and associated artifacts into a box and carried everything back to the Bouyssonie home in La Raufie, where they began to consider where they ought to send the bones for analysis. The Bouyssonie brothers were experts in artifacts, not in skeletal morphology and anatomical descriptions. Like Johann

Fuhlrott fifty years prior, they realized they required the help of an expert in anatomy and taxonomy to make sense of their discovery.

That very night after returning home, August 3, 1908, Jean and Amédée Bouyssonie wrote to two eminent scholars—the renowned French prehistorian Henri Breuil in Paris, as well as Émile Cartailhac in Toulouse—to ask their recommendations for other experts who could provide technical descriptions of the skeleton's anatomy. Breuil was a powerhouse in French prehistory circles, a specialist in geology, prehistory, and ethnography. Cartailhac was best known for his descriptions of the famous Altamira cave paintings in Spain, descriptions that he had completed with Breuil in 1880. Breuil wrote back and suggested that they contact the prominent geologist and paleontologist Marcellin Boule, director of the prestigious Muséum National d'Histoire Naturelle in Paris.

Boule's reputation in the area of human evolutionary research was legendary and his interest in a find as spectacular as the remains from La Chapelle-aux-Saints would have been unmistakable. Boule's own research and work with fossils and geology ranged from Europe to the Middle East, including northern Africa, and his expertise was in correlating sites with geological strata; this meant that Boule was able to establish finds within their proper geologic timeline. Boule was also committed to the processes of disseminating science and information—he was the editor of *L'Anthropologie* for forty-seven years, until 1940. When Boule received the Bouyssonies' letter in 1908, he immediately agreed to study the La Chapelle skeleton, which would arrive at his museum laboratory in early 1909.

The question of where to send the skeleton for analysis isn't and wasn't as straightforward as one might imagine. While the Bouyssonies were confident that Boule's interest and expertise in prehistory and anatomy would offer a scientifically valid interpretation of their discovery, Boule's institutional association with the Muséum National d'Histoire Naturelle played a significant factor. In early-twentieth-century France, archaeology and natural history shared strong historical ties with the theology of the Catholic Church—a relationship typical of the nineteenth century. The question of whom to send it to and where it would be studied had political as well as archaeological significance. Since the skeleton was recovered near the village church—and the excavators were respected clerics—all of the subsequent discussions about where to actually send the bones were mediated through church connections. The other possible destination for the skeleton, the École d'Anthropologie, was simply less appealing to the clerics; this was mostly due to the École's radical politics and its commitment to a philosophy of materialism, to say nothing of its anticlericalism. The École's former director Adrien de Mortillet had argued that "the basis of the universal law of morphological and cultural progress, paleoanthropology and Paleolithic archaeology were political weapons for radical socialist aims, with human history as an integral part and logical consequence of human prehistory."[9] This was a political position rather off-putting to the abbés Bouyssonies, to say the least.

While the École d'Anthropologie might have had the scientific expertise to examine the Neanderthal from La Chapelle, it lacked the standing between science and prehistoric research acceptable to the Catholic Church. Thus the École's loss was the Muséum's gain and the skeleton went to Boule.

———

For the next two years, Boule analyzed, sketched, and studied the Old Man's skeleton, and his 1911 publication, *L'Homme Fossile de La Chapelle-aux-Saints*, was a masterpiece. The monograph was a complete summary of the Old Man, beginning with the fossil's excavation and ending with comparisons to other Neanderthal specimens across Europe. *L'Homme* was filled with chapters of anatomical descriptions, careful measurements, and photographs of the specimen as well as the site of La Chapelle itself.

Each chapter of *L'Homme* showed tables of careful measurements and comparisons with other Neanderthals (most comparisons were made with the Neanderthal from Spy, Belgium, discovered in 1886) as well as other great ape populations. Under Boule's direction, Monsieur J. Papoint, of the Laboratoire de Paléontologie at the Muséum National d'Histoire Naturelle, contributed dozens of pen-and-ink sketches and photography to *L'Homme*; these sketches were anatomical comparisons between the Old Man and modern humans, in addition to drawings of the stone tools found in the La Chapelle site's original excavations.

The book also contained sixteen beautifully detailed stereoscopic reprints of each bone from the skeleton—the 1911 version of 3-D data sharing. The stereoscope was an important tool for laboratory and scientific work in the late nineteenth and early twentieth centuries for a variety of scientific disciplines, including paleoanthropology. The stereoscope expanded what researchers were able to "see" and how they were able to see it, in the same way that telescopes and microscopes expanded the visual possibilities for other sciences centuries before. A stereoscopic plate contains two slightly offset views of the same image, and these images line up with the viewer's left and right eyes. Thanks to the power of binocular vision, the brain "combines" these two images into one, creating the illusion of three-dimensional depth.[10]

At 278 pages, Boule's work was comprehensive, his comparisons thoughtful, and his research judiciously in keeping with other then contemporary tomes of prehistory and anatomy. Because *L'Homme Fossile de La Chapelle-aux-Saints* was the first and most comprehensive publication of Neanderthals in scientific literature, it established the La Chapelle skeleton as the most complete reference for new Neanderthal fossils discovered—thanks in no small part to Boule's detailed studies. Although the 1856 Neanderthal 1 from Germany was the species type specimen—the fossil that researchers had designated as best defining Neanderthals—the La Chapelle skeleton quickly became the go-to fossil for researchers.

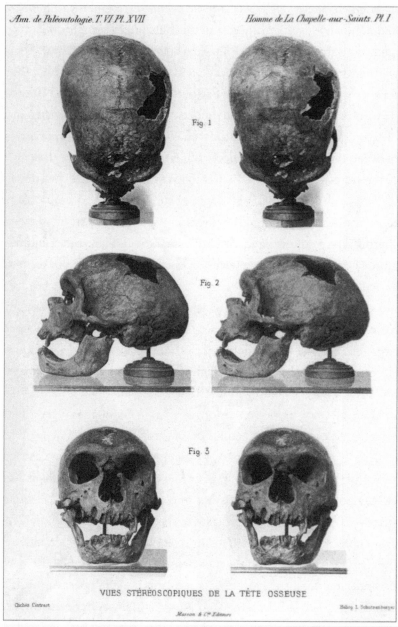

Ann. de Paléontologie. T. VI. Pl. XVII   *Homme de La Chapelle-aux-Saints. Pl. I*

Fig. 1

Fig. 2

Fig. 3

VUES STÉRÉOSCOPIQUES DE LA TÊTE OSSEUSE

Clichés Cintract   Masson & Cie Éditeurs   Héliog. L. Schutzenberger

Stereoscopic images (above and opposite) allowed readers to "see" the subject in 3-D; for the La Chapelle Neanderthal, this meant that readers could "view"

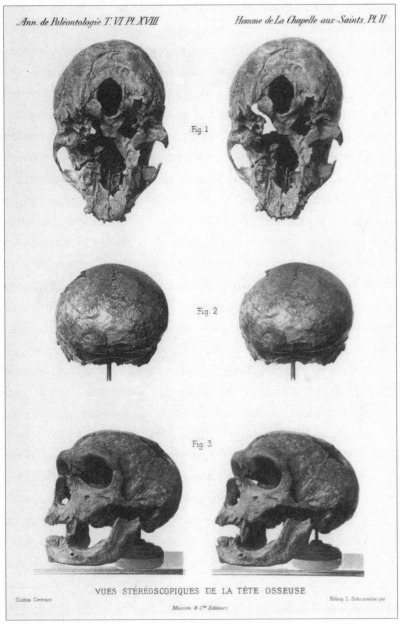

the skull without needing access to a cast of the fossil. Printed in Marcellin Boule's *L'Homme Fossile de La Chapelle-aux-Saints*, 1911.

While Boule worked on his magnum opus, archaeological excavations in France accelerated. Prehistorians realized the archaeological potential of the caves and were quick to begin subsequent excavations in the area. By 1911, three years after the excavations of La Chapelle, the sites of Le Moustier, La Ferrassie, and Cap Blanc had all been excavated by a variety of research teams, and several prehistoric skeletons had been recovered. (Boule actually used parts of a La Ferrassie Neanderthal skeleton, excavated between 1909 and 1911, to fill in missing pieces to the La Chapelle specimen.) Some of these post–La Chapelle skeletons were immediately classified as Neanderthal using the guidelines that Boule had pioneered, while others, like Cap Blanc, were a bit more tricky to sort out taxonomically. With Boule's detailed anatomical and cultural assessment of La Chapelle's Old Man, there was a framework for the newer, post–La Chapelle skeletons. Since Boule's descriptions and reconstructions of La Chapelle became the basis for any and all Neanderthal research that was conducted from the early to mid-twentieth century, his conclusions went unchallenged for decades.

Just what were Boule's conclusions about the Old Man? And how did *L'Homme de La Chapelle-aux-Saints* describe Neanderthals as a species? According to Boule, the Old Man was a truly sad specimen of nature. He was unable to walk upright properly and certainly wouldn't have been capable of any kind of complex behavior or cultural sophistication. Boule reconstructed this skeleton with a severely curved spine, giving the Neanderthal a stooped, slouching stance; he gave the Old Man bent knees and a head that jutted forward. Boule believed that the low-vaulted cranium (the oblong shape that had earlier intrigued Fuhlrott and Schaaffhausen in the 1850s) and the large brow ridge were indications that the crania and the brain it encased were primitive and not as advanced as *Homo sapiens*, meaning that early humans would have lacked intelligence and cultural sophistication—a rather comfortable explanation for Neanderthals' evolutionary demise. Boule gave his reconstruction an opposable big toe like the great apes, even though there was no compelling reason to justify this interpretation and it was simply one more anatomical characteristic that made Neanderthals less like *Homo sapiens*. In short, Boule described the Old Man as what we might see today as a kind of archetypical caveman— not a charismatic Fred Flintstone but a savage, shuffling troglodyte bumbling his way across glaciated Europe.

Boule's reconstruction and reading of Neanderthals had the ability to appeal to a variety of scientists at the beginning of the twentieth century. Boule presented the Old Man as a missing link in the chain of human evolution (pointing to key features of the fossil's anatomy), but he refused to use the species as an intermediate between apes and humans. This particular view of human evolution meant that Boule did not necessarily think that fossil species operated by a strict unilinear model, where every extinct fossil species would neatly line up according to ancestral relationships, ultimately ending with *Homo sapiens*. Boule's evolutionary model allowed for philosophical and taxonomic flexibility in how researchers could think about models of evolution. This approach allowed that—no matter what evolutionary model a researcher ascribed to—Neanderthals could and ought to have a place in it, and Boule's careful study certainly set expectations and grounded interpretations of the Neanderthal species. Any subsequent studies of the Neanderthal skeleton would have to intellectually deal with Boule and Boule's original interpretation as well as any new Neanderthal bones. Boule's work became a "type" of sorts— a type case for how to complete a thorough examination of skeletal anatomy, for negotiating tricky parts of evolutionary theory, and for folding a fossil into a broader milieu of science and popular culture.

Back in the Dordogne region of France, Amédée Bouyssonie, Jean Bouyssonie, and Louis Bardon published the results of their archaeological excavations at La Chapelle-aux-Saints in *L'Anthropologie* in 1909. As a result of their incredibly detailed excavations in the region, they collected over one thousand artifacts from the La Chapelle *bouffia*. In addition to the Neanderthal skeleton, other mammalian bones were recovered in the cave: rhinoceros, horse, wild boar, bison, hyena, and wolf, not to mention a plethora of stone tool artifacts. After Boule concluded his study of the Old Man, the Muséum National d'Histoire Naturelle in Paris purchased the remains for 1,500 francs in 1911. The two Bouyssonie brothers went back to their excavations in the *grottes*—*bouffias*—in south-central France and continued to contribute monographs and manuscripts to the Société Préhistorique Française through the 1950s.[11]

Photograph of the Neanderthal's skull in situ before its removal during excavations at La Chapelle. Published in *Cosmos*, July 1909; similar photographs were published in Boule's *L'Homme Fossile de La Chapelle-aux-Saints*, 1911.

So what did the Old Man really bring to the scientific table? It was an entire specimen. It was in a burial context. The fossil was carefully excavated in situ—none of the other Neanderthals could really be considered as having been collected through archaeological excavations. It had been excavated by experts, studied by experts, and entered the scientific literature through "proper" channels. The Old

Man's credentials, as it were, were impeccable. La Chapelle was the right type of find in the right place at the right time—a celebrated specimen that could preside over the nascent discipline of paleoanthropology. Even more than the Neanderthal species type specimen, Neanderthal 1 from Germany, the Old Man of La Chapelle became the referential type specimen—the archetype—for what scientific and public audiences expected from Neanderthals.

———

The Old Man created a cacophony of publicity for the Neanderthal species. Early twentieth-century artists, scientists, and media formed a triumvirate that promoted interest in Neanderthals. The Old Man also caught the public imagination, thanks in large part to a series of newspaper articles that had some serious academic and popular clout. (Boule himself collected press clippings that described his work with this specimen.) Due to changes in both newspaper printing technology as well as the advent of compulsory elementary education, articles published in print had the potential to reach a large number of readers—the belle époque of French print journalism—and those many readers could and would meet the Old Man. The entire status of a newspaper article underwent a rapid transformation, legitimizing it as a media source and, in turn, sanctioning the subjects it published.[12]

The press wasn't just a tool of cultivated classes anymore; it had clout and power and offered scientists who published in the paper considerable fame and political influence. Photos and reconstructions of the La Chapelle-aux-Saints Neanderthal published in newspapers like *L'Illustration*, the *Illustrated London News*, and *Harper's Weekly* served as go-betweens for "academics" and "the public"—translating scientific ideas to multiple audiences. The Old Man became immensely popular and publically recognizable through Frantisek Kupka's sensationalized art of the *Illustrated London News*—where a hairy apelike creature crawled along cave walls. Paintings by the eminent paleoartist Charles Knight in the early 1900s, however, offered a more nuanced and reflective take on Neanderthals. Knight's paintings in the American Museum of Natural History offered museumgoers a rather empathetic slice of Neanderthal evolutionary history, showing social groups with hunting technology.[13]

Neanderthal reconstruction by F. Kupka, *Illustrated London News*, 1909. This reconstruction has practically become iconic as an example of the bias of early-twentieth-century understanding of Neanderthals.

As Boule's science began to work its way out of the lab and into newspaper articles and museums, Neanderthals began to pique public interest in other ways, particularly through literature. The Old Man found an interesting place in popular imagination thanks in large part to the fledgling genre of science fiction. In the imagined science fiction worlds of Jules Verne, H. G. Wells, and other early sci-fi authors, unexplored geographies, Darwinism, and mechanical inventions populated their alternative worlds. For the Belgian writers and brothers Joseph Henri Honoré Boex and Séraphin Justin François Boex, the caves and archaeological sites of Europe provided a perfect backdrop for imagined, speculative histories, and Neanderthals were a perfect species to populate them. Once the Old Man moved out of Boule's laboratory, the Boex brothers wrote him into a sci-fi novella, giving the Old Man a popular, culturally recursive dimension that Boule could only dream of.

The two brothers published *La Guerre du Feu*—translated into English years later as *The Quest for Fire*—under the nom de plume J. H. Rosny in 1911, the same year as Boule's *L'Homme Fossile de La Chapelle-aux-Saints*. (Short articles about the Old Man had been published between 1908 and 1911, providing the Boex brothers with sufficient Neanderthal source material for their novella.) Science fiction is a powerful genre to explore a certain type of history of science, through speculation about possible scenarios. It takes a discovery (like a fossil species) and asks, "What if?" And it's this what-if that invests the audience in the Neanderthals. For the brothers, Neanderthals were more than just the sum of their archaeological artifacts. Neanderthals were characters who could have motivations and desires, agency and history.

The Quest for Fire is set in the late Paleolithic, where several hominin groups vie for the mastery of fire. Characters who can master, control, and—most significantly—make fire will be the ones who emerge as evolutionary successes. By 1909, there were three major types of fossil species agreed upon in scientific literature—*Pithecanthropus* (fossils from Java, in Southeast Asia, which we call today *Homo erectus*), Neanderthals, and "ancient humans," a hodgepodge grouping of fossils loosely attributed to very old *Homo sapiens*. The authors build a cultural progression, pitting "savagery" against "civilization." Tribes of these long-extinct species would need to master cultural behavior to be evolutionary successes. The crux of the human condition lay in the ability to create fire from flint tools.[14]

In the novella, the Neanderthals lose the stored embers of their fire due to an attack by a barbaric tribe—*Homo erectus*. In a key opening scene, after the Oulhamrs (Neanderthals) have lost their fire, the tribe's leader, Faouhm, raises his arms toward the sky and yells, "What will become of the Oulhamrs without Fire? How shall they live on the savanna and in the forest? Who will defend them against shadows and winter blasts? They will have to eat raw meat and bitter plants, never to warm their limbs, and their spearheads will remain soft. The lion, the saber-toothed tiger, the bear, the tiger, the giant hyena will eat them alive during the night. Who will recapture Fire?"[15]

The novella was made into a now cult classic film in 1981. In the film, as in the novella, once fire is lost, it has to be gathered. Humans— not Neanderthals—are the only ones with the ingenuity and ability to master fire. Most significant, however, is an appeal to an innate ingenuity that surrounds fire—to create it, to care for it, and to anthropomorphize it. (A colleague of mine summarized *The Quest for Fire* as "Ron Perlman goes camping. For two hours. With no dialogue.")

Joseph and Séraphin Boex were fascinated by why *Homo sapiens* survived and Neanderthals did not: What gave one species an evolutionary edge for success and not the other? For the Boexes, the answer for humans' success lay squarely with their technological and cognitive superiority—humans had the tools and the smarts to succeed where Neanderthals did not. Today, over one hundred years after the Boex brothers published their novella, archaeological research offers a very different interpretation of Neanderthal life. Archaeologists consider Neanderthals just as smart, just as technologically apt, just as culturally nuanced as *Homo sapiens*. Despite this shift in interpretation, the Neanderthal-as-hapless-caveman motif is so firmly entrenched in our cultural consciousness that it is exceptionally difficult to dislodge it.

Boule's work might have given the original *Quest for Fire* scientific credibility when it was first published, but the novella has given a story and life to Boule's interpretation of the Old Man that endures much longer than any nonfiction about the fossil.[16]

———

After his discovery, description, and immortalization in fiction, the Old Man's own story, as it turns out, was far from over. Well into the mid-twentieth century, scientists would continue to examine and reexamine the Old Man's skeleton as well as his burial context. In the initial excavations and reports, Jean Bouyssonie had noted that the origin of the pit was not natural. In other words, Bouyssonie believed that other Neanderthals in the Old Man's social group had purposefully dug the pit in the cave's floor. For the Bouyssonies as well as Boule, the "unnatural origin" meant that the site was a burial, with the Old Man's corpse deliberately placed there.

In his written descriptions, Bouyssonie included photos of the cave's entrance and the village of La Chapelle itself. In addition to the material cultural remains—the stone tools and mammal bones—that he describes as part of the excavation, he also validates the cave's geologic and stratigraphic integrity, confirming the value of the in situ artifact and pointing toward French geologist Pierre Martel's involvement with the project. Martel examined the sedimentary sequences and the different strata of the site immediately after excavation. He argued that the shape and origination of the pit could not have been formed from runoff or any kind of scouring action. The pit was angled southeast by northwest and was roughly rectangular in shape. It sat in the middle of the cave approximately one meter from the cave's back wall and was buried little over a meter below the surface. The cross sections of the pit's excavation showed repeated rockfall from the roof of the cave, where large pieces were mixed in with other sediments. The pit itself, about half a meter deep, had clearly been cut into the underlying bedrock or cave surface—the body and the pit did represent a burial.[17]

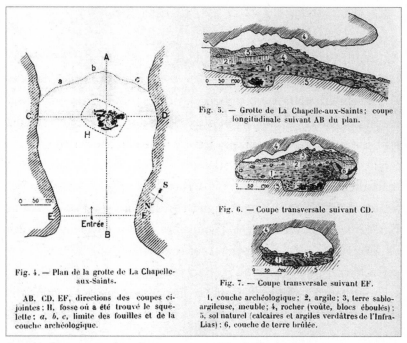

Fig. 4. — Plan de la grotte de La Chapelle-aux-Saints.

AB, CD, EF, directions des coupes ci-jointes : H, fosse où a été trouvé le squé-lette : *a, b, c*, limite des fouilles et de la couche archéologique.

Fig. 5. — Grotte de La Chapelle-aux-Saints; coupe longitudinale suivant AB du plan.

Fig. 6. — Coupe transversale suivant CD.

Fig. 7. — Coupe transversale suivant EF.

1, couche archéologique ; 2, argile ; 3, terre sablo-argileuse, meuble ; 4, rocher (voûte, blocs éboulés) ; 5, sol naturel (calcaires et argiles verdâtres de l'Infra-Lias) ; 6, couche de terre brûlée.

Amédée Bouyssonie, Jean Bouyssonie, and Louis Bardon began excavations at La Chapelle-aux-Saints in 1908. The map of their excavations shows where the Neanderthal skeleton was discovered. Printed in Marcellin Boule's *L'Homme Fossile de La Chapelle-aux-Saints*, 1911.

The idea of a nonhuman species burying its dead challenged notions of behavior and definitions of humanity. If, as the rhetoric went, "human" culture gave us some sort of evolutionary edge, it was awfully uncomfortable to think about a "failed" species—like Neanderthals—as having culture as well. This tension between humans and Neanderthals—almost a moral judgment against Neanderthals—has underscored our perceptions of Neanderthals for well over one hundred years.

By the 1950s, anthropologists began to rework Boule's conclusions about the Old Man's anatomy and culture—perhaps, anthropologists argued, he wasn't as hapless as one thought. This reevaluation focused in two directions: first, reanalyzing the morphology of the La Chapelle skeleton, and, second, reassessing the cultural implications of deliberate burial within Neanderthal culture.

In Paris, on July 25, 1955, anthropologists William Straus and A. J. E. Cave began a major reevaluation of the La Chapelle skeleton, thanks to an impromptu trip to the Musée de l'Homme in Paris while attending the Sixth International Anatomical Congress. Originally interested in examining a particularly controversial specimen, the Fontéchevade skull, they found that they would have access only to casts of the materials, which they weren't allowed to take measurements on.[18] Unable to properly examine the Fontéchevade material, Straus and Cave turned their attention and curiosity to the La Chapelle skeleton housed in the museum, which had been made available to them by the museum curator, Mademoiselle L'Estrange.

What Straus and Cave saw shocked them. "We were not prepared for the severity of the osteoarthritis deformities affecting the vertebral column. It soon became clear why, in his reconstruction of the La Chapelle individual, Boule had found it necessary to turn to the Spy, La Ferrassie, and other Neanderthal skeletons for aid."[19] In other words, the patchy, tattered nature of the La Chapelle fossils, combined with the severe pathologies on the bones, made reconstructing *that particular* specimen difficult. A Neanderthal individual like the Old Man was better understood as a conglomeration of other skeletons—skeletons that "filled in" La Chapelle's missing parts.

Straus and Cave proposed to reexamine, remeasure, and reinterpret the Old Man's skeleton. This would help evaluate the legitimacy and accuracy of Boule's reconstructions of Neanderthal poise, posture, and pathology. In addition to questions of Neanderthal biology and culture, their study also spoke to the burgeoning skeletal population of

Neanderthals that had entered the research record. Where La Chapelle was the first mostly complete Neanderthal skeleton recovered (followed quickly by the La Ferrassie remains in 1909), by 1955 many mostly complete Neanderthals were available for study—Skhul and Tabun (discovered between 1929 and 1935 and today described as found in Israel but noted as Palestine in the midcentury literature), Shanidar Valley in Iraq, and Teshik-Tash in Uzbekistan (noted as Russia). And, of course, British archaeologist Dorothy Garrod's discovery of the Neanderthal child "Abel" in 1939 in Gibraltar.[20]

One of the first of Boule's conclusions Straus and Cave reexamined was how the Old Man stood and how he would have walked. Boule described him as having a slouching, stooping posture. Straus and Cave argued that the Old Man suffered from "deforming osteoarthritis"—a condition of that individual's health rather than a species characteristic.[21] Straus and Cave's own studies suggested that the Old Man, despite his osteoarthritis, would have stood much taller, his shoulders squared back, his head in line with the vertebral column. In short, La Chapelle would have had a much different poise and posture than the Boule model allowed.

But a different pose implied a different culture as well. How would La Chapelle—and other "classic Neanderthals"—have behaved? Would we recognize their culture and its practices? Such questions opened a range of new studies about vocalization, cultural exchange, and altruism within Neanderthal culture. Paleoanthropologists and archaeologists began to reexplore whether Neanderthals had any of those characteristics; if so, how would they be expressed in Neanderthal skeletal morphology or in the archaeological record? These new studies indicated that many of the features scientists saw as "uniquely Neanderthal" characteristics actually fell within the range of modern human variation—Neanderthals weren't so different from humans after all.

When questions of Neanderthal altruism—how much a Neanderthal group would have helped its members—emerged, studies routinely pointed back to the La Chapelle specimen. In the 1980s, studies like N. C. Tappen's study of the Old Man's dentition raised serious concerns about the legitimacy of interpreting La Chapelle as the recipient of help from his social group, claiming that the skeleton's dentition was "not reliable evidence of altruistic behavior by his [La Chappelle's] cohort."[22]

An even more recent evaluation of the entire skeleton by scientist Dr. Erik Trinkaus has shown that while the Old Man of La Chapelle did suffer from a degenerative joint disease, the deformation caused by this should not have affected Boule's original reconstruction of the individual's posture. It appears that Boule's own preconceptions about early humans, and his rejection of the hypothesis that Neanderthals were part of *Homo sapiens*' evolutionary cohort, led him to reconstruct a stooped, brutish creature, effectively placing Neanderthals on a side branch of the human evolutionary tree.

"The La Chapelle remains are often held up as a 'type' specimen of Würm glaciation European populations that have come to be known as 'Classic Neanderthals.' One might well inquire into the nature of the 'type' specimen," paleoanthropologists Ronald Carlisle and Michael Siegel noted. "It is almost certainly for reasons of skeletal completeness that La Chapelle has been designated a type, yet it is distinctly an archetype and says nothing by itself of the range of variation that existed within the population of which it was a member."[23] In other words, we have made La Chapelle into an archetype.

By 2014, paleoanthropologist Dr. William Rendu summarized the state of La Chapelle research: "He [the Old Man] was quite old by the time he died, as bone had re-grown along the gums where he had lost several teeth, perhaps decades before. He lacked so many teeth in fact that it's *possible* he needed his food ground down before he was able to eat it. Other Neanderthals in his social group may have supported him in his final years. Finally, the discovery of skeletal elements belonging to the original La Chapelle aux Saints 1 individual, two additional young individuals, and a second adult in the bouffia Bonneval highlights a more complex site-formation history than previously proposed."[24] Why do we keep coming back, then, to the La Chapelle specimen as the archetype for Neanderthal? The fossil's scientific and cultural cachet transcended its existence as simply a comparative specimen in a museum's collection into a fossil with a story, a history, and an identity.

———

As more and more Neanderthal fossils were discovered and published throughout the twentieth and twenty-first centuries, questions about who they were, how they lived, and what the implications were of such a human-but-not-quite-human species began to percolate through a variety of media, from books to museums. Just as the science of Neanderthals continued to evolve and change, so did representations of Neanderthals in museums, literature, and popular culture in the early twentieth century.

Early museum dioramas of the Stone Age traded in the many cultural memes that surrounded Neanderthals. Dioramas and reconstructions became a way to put a body on a fossil; a reconstruction of a fossil provides a visual dimensionality of muscle, skin, hair, and movement that imbues a sense of "realness" to a fossil that a mere skeletal description, however detailed, simply cannot match. For fossils that are unable to freeze life as is (as would, say, an animal preserved by taxidermy), a reconstruction makes extinct species like Neanderthals accessible and understandable. We can immediately imagine the hominin in question—its tangible face and body are right in front of us and the body begs for a story to go with its science.

When we see reconstructed bodies of human ancestors, we summon certain narratives and interpose these underlying motifs onto the figures that we're seeing. More than a plate or a placard, dioramas tell the viewer about a context, a setting, for a story.

Historically, museum dioramas have helped enforce the narrative of Boule's Neanderthal, long popularized by *The Quest for Fire*. Again, this was particularly true at the turn of the twentieth century, when Frantisek Kupka popularized a primitive, hairy Neanderthal with his sketches in the *Illustrated London News*. Museum dioramas of Neanderthals conveyed particular prejudices and biases as well as information. Most of all, they told of particular assumptions, carefully tucked away below the surface of the exhibit.

Chicago Field Museum's diorama of "Mousterian Man" (Neanderthals) as part of its Prehistoric Man series, 1930s. This image was taken from the museum's visitors guide (H. Field and B. Laufer, *Prehistoric Man*, Hall of the Stone Age of the Old World, Field Museum of Natural History, Chicago, 1933).

In July 1933, the Chicago Field Museum of Natural History installed eight dioramas that depicted scenes from "early" hominin life from a variety of then current archaeological sites across Europe. These dioramas, by sculptor Frederick Blaschke, typify many early-twentieth-century assumptions about Neanderthal prehistory and the guesswork that was involved in reconstructing a spotty archeo-paleontological record. The portrayal of technology in Blaschke's dioramas, in particular, creates a powerful, indirect argument about success and direction in human evolution.

In Blaschke's imagined scenes, Neanderthal hands clutch at a tool but lack any kind of dexterity—the tools and their use look clumsy. More than just fossils come to life, the slouching, hunched recon-structed Neanderthals reflected an interesting, subtle thesis about material culture and how hominins "ought" to have interacted with

Restoration of a Neanderthal man in profile, Field Museum of Natural His-tory, Chicago, 1930s. This Neanderthal reflects the 1930s interpretations of the species: a hunched back and thick neck, very much in keeping with Marcellin Boule's conclusions. *(Wellcome Library, London; CC-BY-4.0)*

it. The reconstruction assumed a disconnect between tools and tool-makers; Neanderthals simply did not have anything material that was evolutionarily compelling. No complex tools, no dexterity, no ingenuity to invent "good" technology. Neanderthals, as portrayed in dioramas like Blaschke's, solidified a story that champions technological ingenuity for humans' "success." The more people met these visual hominins in the museums, the more the Neanderthal stereotypes became locked in our cultural imagination.

Modern museum exhibits are working to change our cultural assumptions and to give museumgoers a better taste of current archaeological research. Because exhibits and dioramas have such incredible visual staying power, they offer a way to introduce a more nuanced Neanderthal to museumgoers. The Hall of Human Origins at the Smithsonian Institution, for example, offers a glimpse of our changing cultural assumptions about Neanderthals. When the Hall of Human Origins opened in 2010, it combined a display of the actual remains of a Neanderthal from Shanidar, Iraq, with an exhibit where visitors can download an app to "transform themselves into an early human." (Called MEanderthal, the app creates a composite of your own face and an early hominin reconstruction.) Paleoartist John Gurche described the Neanderthal reconstruction that he created for the exhibit as "a behaviorally sophisticated kind of human. . . . I wanted to portray a being with a complex inner life. A distinct hairstyle . . . and a deerskin hair band with a lined design, hint that this complex being has symbolic levels to his thinking."[25] This is a long way from Boule's 1957 description in his popular textbook *Fossil Men*: "There is hardly a more rudimentary or degraded form of industry than our Mousterian [Neanderthal] Man. . . . [T]he brutish appearance of this energetic and clumsy body, of the heavy-jawed skull . . . declares the predominance of functions of a purely vegetative or bestial kind over the functions of mind."[26] The changes to

43

Neanderthal exhibits—"humanizing" them—mean that it is much easier to see the fossil species as someone like you and less like an evolutionary oddity.

Reconstruction of a Neanderthal (*Homo neanderthalensis*) based on the La Chapelle-aux-Saints fossils. Reconstruction by Elisabeth Daynes of the Daynes Studio, Paris, France. *(Sebastien Plailly/Science Source)*

Museums aren't the only means through which Neanderthals' popular image is undergoing rehabilitation here in the twenty-first century. Almost one hundred years after the publication of *The Quest for Fire*, Canadian science fiction author Robert Sawyer imagines an alternative evolutionary timeline where Neanderthals—not humans—were the evolutionary "successes" of the Pleistocene. In his Neanderthal Parallax trilogy, published in 2002–2003 (*Hominids, Humans, Hybrids*), Sawyer proposes: What if *Homo sapiens* hadn't been the only member of the genus *Homo* to survive to the present day? What if that evolutionary history went to some other species, Neanderthals? What if Neanderthals had spent the last thirty thousand years achieving a culture like ours, which we believe to be uniquely "human"? And, most provocatively, what if Neanderthals did "human" better than us?

In the trilogy, Sawyer lays out two different Earths—an Earth as we traditionally consider it, and an Earth where Neanderthals became the dominant hominin 250,000 years prior. In this parallel world it was the humans (or *gliksin*), not Neanderthals, who went extinct. The Neanderthal Earth crosses with ours when Neanderthal physicist Ponter Boddit manages to travel between the two, through a portal that opens up at the Sudbury Neutrino Observatory's particle physics lab.

Back in the real worlds of archaeology and paleoanthropology, scientific consensus puts the extinction of the Neanderthals around thirty thousand years ago. Purported causes have ranged from changing climates and inferior technologies to the advent of revolutionary cognition in *Homo sapiens*—all hypotheses Sawyer works in to his speculative anthropology. Most explanations for Neanderthal extinction hinge on some spark of ingenuity that humans possess and Neanderthals don't. Yet recent studies of Neanderthals offer serious challenges to these long-held views. From Italy, Gibraltar, Portugal, and Spain, such studies show Neanderthals as complex hominins, capable of sophisticated behavior—capable, indeed, of behavior generally thought to be exclusive to *Homo sapiens*.

The winner of the World Science Fiction Society's prestigious

Hugo Award, *Hominids* examines the interactions and moral implications of Neanderthals and humans through interspecies relationships. Sawyer's attention to detail and his paleoanthropological research provide the same gestures toward scientific legitimacy as Rosny's survey of La Chapelle. Sawyer's details—just like Rosny's—are well researched and ring just true enough to lend anthropological legitimacy to the stories. (In a fantastic literary twist, even the names of the human protagonists—Louise and Mary—are gestures toward paleoanthropology's history, acknowledging paleoanthropologists Louise and Mary Leakey.)

———

It's hard—if not impossible—to crack open a book about human evolution and not read about the history and significance of Neanderthals. They remain the first discovered fossil hominin species, and over the last 150 years they have given us a framework to think about them as a character, a species, and a concept.

Neanderthals have been written into the evolutionary narrative as an "Other"—a foil, a double, an easy contrast to our culture that discovered and interpreted them. As any student of literature will tell you, a foil takes a particular character and heightens that character by comparing him with another one; the foil is usually created to project the protagonist. Sherlock Holmes and Dr. Watson. Don Quixote and Sancho Panza. Dr. Jekyll and Mr. Hyde. The most effective foils are generally created by contrasting the two through some set of essential characteristics.

In order for a foil to be truly successful, the character must have something in common with the story's protagonist. Twentieth-century writer Vladimir Nabokov imagined Shakespeare's Caliban and Ariel as classic foils that illustrated the opposite directions of the human condition. Caliban, the savage: "You taught me language; and my profit on't / Is, I know how to curse. / The red plague rid you / For learning me your language!" Ariel, the ethereal spirit: "Pardon, master; I will be correspondent to command / And do my spiriting gently."[27] Ariel entreats and Caliban practically growls—civilization and barbarity are neatly juxtaposed. Four hundred years after Shakespeare penned *The Tempest*, Caliban's savage caricature still serves, for better or worse, as the central trope in our understanding of Neanderthals; we can think of them only when we first cast ourselves as protagonists in the narrative of human evolution. (Sawyer's science fiction trilogy asks its audience to consider which—Neanderthals or humans—is the real Caliban character. In other words, which character is "more human"? In the trilogy, Caliban switches places with Ariel—the wise fool who teaches humanity how to be human.)

Today, the Old Man's bones reside in the Muséum National d'Histoire Naturelle in Paris, and images of the fossil show up in scientific studies and popular museum exhibits. Over the last one hundred years, the Old Man and his Neanderthal contemporaries have undergone serious changes in the definition of Neanderthals as a species—our own notions of humanness challenge our notions of Neanderthalness. These changes in Neanderthal research and studies have unfolded in archaeology, paleoanthropology, genetics, and museum theory (how the species is displayed to audiences). Somewhere in the century since the fossil's discovery, the Old Man moved from an "it" to a "him." He has a personality and a temperament. He also has a purpose.

The discovery of La Chapelle-aux-Saints 1, the Old Man, required a particular framework to truly make sense of what a human-but-not-human fossil species meant. This extended beyond simply accepting evolution as a mechanism of change or the legitimacy of the Neanderthals as a separate species. It necessitated a cultural component—a metaphor or archetype—that was easily accessible in culture writ large. This framework came from other cultural tropes and analogies, mechanisms that allowed culture and science to seamlessly intersect and offer explanations about a species as curious as the Neanderthals.

The history of Neanderthals' discovery has been told many times and in many ways. Where many might frame this history through the Neanderthals' interpretation as a "missing link," many other explanatory devices, including literature, can tell us more about how the species was internalized and used. Indeed, the idea that nineteenth-century science would reach for analogies and metaphors to explain fossil discoveries isn't far-fetched. That it should look for such explanations in literary characters and tropes shows how much science drew from literature during this time. This meant being able to explain Neanderthals beyond simply their evolutionary mechanisms—they had to make sense culturally as well. And while the La Chapelle skeleton might not offer an explanation for all Neanderthal behavior, any explanation of Neanderthals in scientific literature or popular media deals, by necessity, with the legacy of La Chapelle's interpretations.

Today, the Old Man's fame comes from a curious mix of science, history, and even caricature—he is a phylogenetic foil for *Homo sapiens*. "We see ourselves, for better or worse, in comparison to Neanderthals," suggests archaeologist Dr. Julien Riel-Salvatore. "We want to see how we stand out, but lately research seems to shy away from the question of direct competition with *Homo sapiens*. We're moving toward a more nuanced understanding of the species; offering hypotheses that are not just quasi-biological explanations for human-Neanderthal interactions or a strict biological determinism to explain Neanderthal extinction."[28]

Today, the Old Man is more than just the sum of his studies—more than just his skeleton and more than simple scientific evidence.

After his discovery, he became a character *de force* in hominin evolutionary history. Like a dignified family patriarch, the Old Man presides over our evolutionary story. He is paleoanthropology's first famous fossil and he continues to resonate in both the scientific and popular imagination.

Charles Dawson holding cast of Piltdown skull, ca. 1914.
(The Trustees of the Natural History Museum, London. Used with permission)

CHAPTER TWO

PILTDOWN: A NAME WITHOUT A FOSSIL

On February 14, 1912, Charles Dawson, a legal solicitor and well-known artifact collector, happened to notice some peculiarities about a bed of gravel that contained bits of bone near Barkham Manor, at Piltdown, close to his hometown of Lewes in southern England. Dawson's curiosity about the site prompted him to write to his friend and colleague Arthur Smith Woodward, keeper of geology at the British Museum (Natural History) in London, to tell him of his discovery.

"I have come across a very old Pleistocene (?) bed overlying the Hastings Bed between Uckfield and Crowborough which I think is going to be interesting," Dawson penned to Smith Woodward in a letter dated the following day. "It has a lot of iron-stained flint in it, so I suppose it is the oldest known flint gravel in the Weald. I (think) portion of a human (?) skull [*sic*] which will rival *H. Heidelbergensis* [*sic*] in solidarity."[1]

Dawson was the author of the extensive two-volume *History of Hastings Castle* and an avowed antiquarian enthusiast. He was a fellow of the Society of Antiquaries and of the Geological Society of

London. For years, Dawson had collected fossils around the Lewes area and sent them to Smith Woodward and the British Museum. But the discovery of these flint artifacts and the portion of a human skull mentioned in that February letter were curiosities of a different kind. While Paleolithic Stone Age materials had been found in England for decades by amateurs and professional collectors alike, these artifacts had not been found with bones old enough to be a species different and older than *Homo sapiens.*

The Heidelberg fossil from Germany that Dawson referenced in his letter was a humanlike jaw discovered in 1907—the Mauer mandible, as it became known—that had pushed back the age of humanity's antiquity in Europe. The Mauer mandible was significant to researchers, like Smith Woodward and Dawson, interested in *Homo sapiens* antiquity in Europe. Alluding to this fossil, naturally, would have piqued Smith Woodward's interest. The discovery of an ancient human from the Pleistocene was something completely new for the British paleointelligentsia. While Britain did have evidence of ancient humans—from the geologically recent Holocene—nothing as old as the Pleistocene had been found prior to Dawson's discovery. Given this, Arthur Smith Woodward was more than just a little interested in learning about the fossil remains, and so he set about organizing further excavations of the Barkham Manor gravel pit.

The turn of the twentieth century was an exciting time for paleontological discoveries, and fossils captured scientific and public imaginations alike. In that original letter to Smith Woodward, Dawson also wrote, "Yes, Conan Doyle is writing a sort of Jules Verne book on some wonderful plateau in S. America with a lake which somehow got isolation from 'Oolitic' times and contained old the [sic] fauna and flora of that period, and was visited by the unusual 'Professor.' I hope someone has sorted out his fossils for him!"[2] Arthur Conan Doyle's issues with South American fossils aside, little could either Dawson or Smith Woodward imagine that the portion of human skull from those very old Pleistocene beds at Piltdown near East Sussex would easily be the most famous—or infamous—discovery in the history of studying human evolution. Over the last hundred years, the legend and mystique of Piltdown has well outgrown its humble gravel origins.

———

When Dawson's discovery at Piltdown entered the paleodiscourse in 1912, it was a truly curious find. To begin with, the fossil's anatomy was a bit of a conundrum—when assembled together, the fossil appeared to have an apelike jaw and humanlike skull, suggesting the fossil could be the perfect "missing link" between apes and humans. The shape and features of the skull seemed to emphasize humans' distinctive "big brains" and suggested that we acquired the capacity for complex thought very early in our evolutionary history. With those characteristics, the fossil lent a credibility, even legitimacy, to a narrative of unilinear evolution—that humans were the culminating end point of primate evolution. Fragmented though it was, the Piltdown fossil provided a very neatly packaged evolutionary story for the antiquity of *Homo sapiens*.

But the irony was that the Piltdown fossil wasn't really a fossil human ancestor—it wasn't even a "real" fossil. In the early 1950s, the Piltdown materials were found to be a hoax. It was fossil forgery of the first degree, comprising real but very recently old human crania, orangutan bones, and chimpanzee teeth all masquerading as a fossil much older than it was. For early twentieth-century studies in human evolution, Piltdown was simply a critical piece of evidence for sorting out humans' ancestral family tree; by midcentury, Piltdown was a social *experimentum crucis*—a litmus test, if you will, for using new technologies and methodologies against long-held beliefs about the specimen as a fossil ancestor. More than one hundred years after its discovery, Piltdown is an unsolved mystery as well as a cautionary tale for bending fact to fit the theory.[3]

Consequently, Piltdown Man remains one of the most-studied but least-resolved fossils within paleoanthropology, and for forty years the fossil has been an anchoring point for interpreting hominins and hominin phylogeny. But why? And how? How did the fossil move from a "portion of a human (?) skull [*sic*] which will rival *H. Heidelbergensis* [*sic*] in solidarity" to a scientific and social problem that needed a solution? Why does this problematic fossil have an incredible staying force within the scientific field even today?

In the first decade of the twentieth century, the fledgling discipline of paleoanthropology had precious few fossils to hang their science on. It could boast a couple of Neanderthal skulls from France, the Old Man's skeleton, of course, the *Homo heidelbergensis* jaw from Germany, some other scattered skeletal elements from around Europe, a skull from Australia, and a few other bits and fragments here and there. The reigning fossil du jour was the Dutch anatomist Eugène Dubois's 1891 discovery of Java Man (which Dubois termed *Pithecanthropus erectus*), found in Trinil, Indonesia, which dominated the paleointellectual landscape for decades.

Although many Paleolithic stone tool artifacts had been found in Britain by the early twentieth century, there wasn't any kind of skeletal candidate for human ancestry that indicated the geologic antiquity of early *Homo sapiens* in Britain. If the stone tools and other evidence of early man were being recovered, the logic went, then it should be only a matter of time until a suitable skeleton (from the geologically old Pleistocene era) was discovered with them. But the question still fronted the scientific community: Where was the elusive skeleton of "early man"—as he was called—in Britain?

———

After the initial discovery of the Piltdown materials and the ensuing February 1912 letter, the fossil and its excavation were kept a closely guarded secret from the media's prying eyes to allow scientists to study the fossil carefully. In his official account of the Piltdown discovery, published in the *Quarterly Journal of the Geological Society of London* later that same year, Charles Dawson suggested that his interest in the Piltdown site had been piqued well before 1912:

> Several years ago I was walking along a farm-road close to Piltdown Common, Fletching (Sussex), when I noticed that the road had been mended with some peculiar brown flints not usual in the district. . . . Upon one of my subsequent visits to the pit, one of the men handed me a small portion of an unusually thick human parietal bone . . .
>
> It was not until some years later, in the autumn of 1911, on a visit to the spot, that I picked up, among the rain-washed spoil-heaps of the gravel-pit, another and larger piece belonging to the frontal region of the same skull, including a portion of the left superciliary ridge . . . I accordingly took it to Dr. A. Smith Woodward at the British Museum (Natural History) for comparison and determination. He was immediately impressed with the importance of the discovery, and we decided to amply labour and to make a systematic search among the spoil-heaps and gravel, as soon as the flood had abated; for the gravel-pit is more or less under water during five or six months of the year. We accordingly gave up as much time as we could spare since last spring (1912), and completely turned over and sifted what spoil-material remained; we also dug up and sifted such portions of the gravel as had been left undisturbed by the workmen.[4]

This account of the fossil's discovery was read at a meeting of the Geological Society of London on December 18, 1912. However, various newspapers that covered that meeting quoted Dawson as claiming to have been first handed a fragment of cranium "four years ago," putting the "discovery" of the fossil in 1908.

Perhaps more spectacularly, Dawson claimed that the Piltdown cranial fragments had been accidently broken and then discarded by the workers at the gravel pit, where Dawson alleged the workers said that the pieces looked like broken "cocoa-nuts." (The original notes to his portion of the Geological Society paper provide us with a bit of archival evidence to the "coconut" story associated with Piltdown.) Under the heading "Brief Story of Discovery" in his *Quarterly Journal* article, Dawson wrote: "Human skull found and broken by workmen. Hence subsequent digging both in spoil-material and in the bottom layer of gravel left untouched by them."[5] In fact, two versions of the "coconut" story appear in the newspaper. The first relates how Dawson was handed a fragment of the broken skull and his subsequent efforts to recover any other discarded pieces. The second description of the coconut story reports that all of the coconut parts of the specimen were discarded and then recounts Dawson's efforts to recover them.[6]

Regardless of the exact circumstances of the fossil's discovery, upon receiving Dawson's letter Smith Woodward agreed to visit the site and considered it worthwhile to launch an excavation and formal investigation. Throughout the summer of 1912—at Smith Woodward's and Dawson's own expense, as Smith Woodward's wife, Lady Maud, recalled years later—the first field season of digging was completed on weekends by a few trusted colleagues. Smith Woodward came down from London and lodged with his wife at the railway hotel in Uckfield or at Dawson's home in Lewes.

In his memoir, *The Earliest Englishman*, Smith Woodward described some hilarity that surrounded those 1912 excavations: "Both the landowner and the farmer had given Mr. Dawson permission to explore the gravel pit at Barkham Manor without knowing precisely what was his object. He had merely expressed interest in the brown flints found there. The eagerness with which we all dug and sifted gravel during the first week therefore excited much interest and curiosity in the neighbourhood."[7] One can almost picture a *Downton Abbey*–esque moment where life at the manor was interrupted by a group of guys digging around the road leading to the property. "The police were informed," Smith Woodward recalled, "and the following Monday morning the local constable appeared at Mr. Dawson's office in Uck-field (where he was Clerk to the Magistrates), stating that he had a report to make. Dr. Dawson, as usual in such cases, admitted the con-stable, and was surprised to learn from him that, 'three toffs, two from London had been digging like mad in the gravelpit at Barkham, and nobody could make out what they were up to.' Mr. Dawson's embar-rassment may easily be imagined, but he remained calm and quietly explained to the constable that there were interesting flints in the neighbourhood, and perhaps the men he reported were merely harm-less seekers after these flints."[8]

The famous French Jesuit prehistorian and philosopher Pierre Teilhard de Chardin, of the Muséum National d'Histoire Naturelle, joined Smith Woodward and Dawson's original excavation team in the spring of 1912. In a letter dated May 18, 1912, Teilhard describes the work at Barkham Manor: "I forgot to tell you that when Dawson came along the last time [April 20, 1912] he appeared with a large carefully wrapped box from which he excitedly drew one third of the skull of the 'Homo Lewensis' found by him during these last years in some alluvia (reposing on Wealdian) near Uckfield. The skull is cer-tainly very curious, of deep chocolate colour and especially of a stupefy-ing thickness (about one centimetre at the thinnest points); unfortunately the characteristic parts, orbits, jaws etc. are missing."[9]

Over that 1912 field season, Dawson, Smith Woodward, and Teil-hard collected skeletal remains, mammalian fauna, and artifact imple-ments. The laborers found an additional seven cranial fragments, the right half of a jaw with two molars in situ, as well as a modest assort-ment of fossil animal bones and stone artifacts. Between Dawson's original collections at Barkham Manor in 1908 and 1912, recovered artifacts included a total of ten bone fragments from a cranium and mandible, ten fragments of fauna (ancient hippopotamus, mastodons, and horses, mainly), and twelve artifacts classified as a variety of Paleolithic scrapers, drills, and other stone tools.[10] The Piltdown skull was an isolated find, yes, in that there weren't other remains of human ancestors in the assemblage or other sites like Piltdown in the area, but the skull and jaw were found in the company of mastodon molars and Paleolithic tools—giving Piltdown Man an archaeologi-cal context and an authenticity derived from associated stone tool artifacts.

For the 1912–1913 Piltdown field seasons, Smith Woodward em-ployed local photographer John Frisby to take pictures of the site and excavations, as well as produce a rather formal portrait of Dawson with the Piltdown fossil. Frisby's photographs show Smith Wood-ward and Dawson excavating at the site, often with unnamed labor-ers. (One of the most well-known photographs has "Chipper the Goose" preening his way across the lower left quadrant of the image.) Hiring a photographer to document the Piltdown area illustrates how important the site was to the Piltdown research team.

Central to the Piltdown story is the field site of Piltdown itself. In the early days of the fossil's discovery, the Piltdown quarry func-tioned as a text to be read, interpreted, and reread as the paleointel-ligentsia quibbled over Piltdown's evolutionary relationship with other fossil hominin finds. Since the site was relatively close to the intellectual metropoles for the study of human evolution—as opposed to fossil hominin sites in Java, South Africa, or even rural France—Piltdown was a physical place for researchers to visit and make sense of. That physicality, coupled with reports and photo-graphs from famous scientists, offered a particular legitimacy for the fossil's initial discovery. The presence of a field site made the fossil real in a way that was difficult to challenge. Photographing the excavations was yet another way of cementing that social legitimacy.

One of Frisby's most interestingly poignant photographs from those early days of Piltdown is his portrait, later printed as a postcard, of Charles Dawson. The portrait shows Dawson in a jacket and waist vest with a pocket watch, sitting in a chair with fossils on a table in front of him. He cradles a reconstructed cast of Piltdown in his left hand as he examines a bit of cranium in his right hand. Trees reflect off the glass doors of the bookshelves in the backdrop. Dawson, who died in 1916, looks every bit the part of a proper fossil collector, intrigued by the paleo remains. The photograph also creates an interesting story arc—the fossils are in bits and then must be reconstructed or mediated by the person, with the construction of each "stage" of the paleo process in his hands. There's a sense that Dawson is negotiating the fossils from indistinguishable bits of rock to a fully recognizable fossil ancestor.

———

However much Smith Woodward and Dawson worked to keep the remarkable find veiled in secrecy and their excavations completely under wraps—again, to allow them time for detailed analyses of the fossils—there were rumors of the "remarkable skull" found at Piltdown circulating in the British media by late September 1912. By mid-November, the story was reported in the national press, and the duo began to make preparations to formally present the fossil to the Geological Society of London.

The night of Wednesday, December 18, 1912, the Geological Society was packed to the gills in anticipation of seeing the Piltdown material in the flesh. (Or bone, rather.) The braincase consisted of four large pieces reconstructed from nine fragments. In addition to the fossils, Smith Woodward also unveiled the first of his reconstructions of the fossil, filling in the missing parts of the hominin's face, crania, and jaw. At the Geological Society presentation, Smith Woodward and Dawson presented the fossil's scientific name, *Eoanthropus dawsoni*—"Dawson's dawn ape," in honor of its discoverer. Many in attendance, such as Smith Woodward, the Honorable Professor of Archaeology William Boyd Dawkins, and those associated with the British Museum (Natural History), were excited about the Piltdown fossil since it so perfectly fit with the in-vogue scientific theory that big brains had a particularly long existence.

Smith Woodward claimed that the find pointed to a "missing link" in the chain of human evolution—a fossil that could be reconstructed as a human ancestor with a large brain evidencing the long significance of *Homo sapiens* culture (assuming a big brain was requisite for complex human culture writ large—language, symbology, and so on).

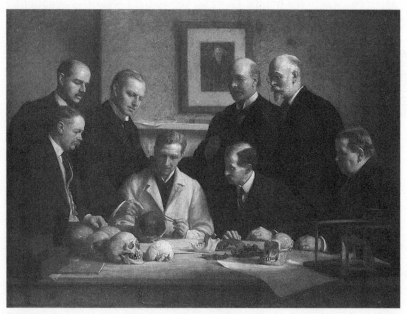

Portrait: *Examining the Piltdown Skull*, by John Cooke, 1915. Back row: F. O. Barlow, G. Elliot Smith, Charles Dawson, Arthur Smith Woodward. Front row: A. S. Underwood, Arthur Keith, W. P. Pycraft, and Ray Lankester. Note Charles Darwin portrait behind the examiners.

Smith Woodward wasn't alone in his interpretation. The Piltdown fossils were folded into the paleo community, and many fossils found in subsequent decades (such as the 1925 Taung Child in South Africa) were ignored due to the wielding influence of Piltdown. Even prominent American paleontologist Henry Fairfield Osborn (then president of the American Museum of Natural History) declared the skull and jaw a perfect fit and the specimen fascinating. In short, the Piltdown fossil offered human evolution a neat narrative with the evidence to back it up. But the complete, unquestioned acceptance of the fossil as a specimen from a single individual of ancient geological age was far from a sure thing, even upon Piltdown's unveiling in 1912.

———

The second Piltdown field season lacked the secrecy that surrounded the first, and in 1913 the Piltdown site was flooded with visitors, especially enthusiasts affiliated with the Geological Association. In fact, going to Piltdown's site became rather like embarking on a holiday excursion for scientists and general public alike. Photographs of that season show the Association's ladies and gentlemen in full Edwardian splendor milling around the site, picnicking, and peering at the excavations.

Archaeologist William Boyd Dawkins, author of the classic 1880 book *Early Man in Britain*, accepted the initial interpretation of Piltdown put forward by Smith Woodward and Dawson. "Man appears in Britain and the Continent at the period when he might be expected to appear," Dawkins argued in *Geology Magazine* in 1915, "from the study of the evolution of the Tertiary Mammalia—at the beginning of the Pleistocene age when the existing Eutherian mammalian species were abundant. He may be looked for in the Pliocene when the existing species were few. In the older strata—Miocene, Oligocene, Eocene—he can only be represented by an ancestry of intermediate forms."[11]

One of the biggest debates in the paleointelligentsia of the early twentieth century was the development of an evolutionary sequence of "humanlike" traits and the order that these traits appear in the fossil record. The question of whether brains developed before or after bipedalism occupied a good proportion of paleo research efforts. The Piltdown fossil seemed to weigh in on all the big questions and was touted as proof that large brains had evolved first.

By 1915, Piltdown was firmly ensconced in the paleo world, despite some detractors. In fact, Piltdown was so completely established in the scientific community, any theory or hypothesis about human evolution had to address the Piltdown fossil, either in support (usually) or as detractors (less so). "The 'dawn man,'" Henry Fairfield Osborn wrote in the 1925 edition of *Men of the Old Stone Age*, referring to *Eoanthropus dawsoni*, "is the most ancient human type in which the form of the head and size of the brain are known. Its anatomy, as well as its geologic antiquity, is therefore of profound interest and worth of very full consideration."[12]

While various scientific communities would continue to debate Piltdown's geological and anatomical details for decades, the fossil came to life through the barrage of newspaper articles and quickly became a staple of museum exhibits on Early Man, thanks to casts and artists' reconstructions of the fossil. Unlike Paleolithic sites in Europe, like La Chapelle, Piltdown was relatively accessible for British scientists interested in examining the fossil's location for themselves, as Barkham Manor was only a train ride away from London.

The British Museum's *Guide to the Fossil Remains of Man* was issued in 1918 specifically to educate visitors about Piltdown. "That he cannot be later than early Pleistocene is proved, if it be admitted that the bone implement shown in fig. 2 (p. 11) was made by Piltdown man; for this implement is fashioned from the middle of the thigh-bone of one of the gigantic elephants (such as *Elephas meridionalis* and *Elephas antiquus*) which lived in Europe in the latter part of the Pliocene and the early part of the Pleistocene period."[13] The hubbub that surrounded Piltdown, where it was discovered, and how it was displayed in museums meant that people were invested in the social success of the fossil.

PREFACE.

MR. CHARLES DAWSON's discovery of the Piltdown skull has aroused so much interest in the study of fossil man, that this small Guide has been prepared to explain its significance. Most of the known specimens important for comparison are represented in the exhibited collection by plaster casts ; and near these, in the same and adjacent cases, are arranged both human implements and associated animal remains to illustrate the circumstances under which early man lived in western Europe.

Thanks are due to the Council of the Geological Society for permission to reproduce Figs. 4 (A, B), 5, 6 (A, B, D), 8-9 (A, B, D), and 12, from the Society's Quarterly Journal.

A. SMITH WOODWARD.

DEPARTMENT OF GEOLOGY,
December, 1914.

P.S.—The only important change in the second edition of this Guide is the addition of the figure and description of a bone implement found in the Piltdown gravel (pp. 11, 12).

A. S. W.

DEPARTMENT OF GEOLOGY,
April, 1918.

The preface to *A Guide to the Fossil Remains of Man*, a pamphlet text for museumgoers, published by the Department of Geology, British Museum (Natural History) (today's Natural History Museum, London), 1918.

Upon the original publication of the Piltdown fossil, Sir Arthur Smith Woodward and Sir Arthur Keith each created a cast, and the reconstructions offered slightly different interpretations of Piltdown's cranial anatomy. When more cranial fragments were recovered during the 1913 field season, scientific consensus favored Smith Woodward's reconstruction over Keith's, implicitly lending credibility to Smith

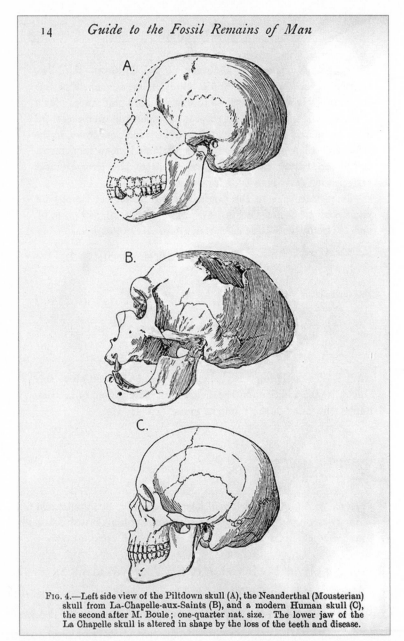

14   *Guide to the Fossil Remains of Man*

A.

B.

C.

FIG. 4.—Left side view of the Piltdown skull (A), the Neanderthal (Mousterian) skull from La-Chapelle-aux-Saints (B), and a modern Human skull (C), the second after M. Boule; one-quarter nat. size. The lower jaw of the La Chapelle skull is altered in shape by the loss of the teeth and disease.

Comparison of La Chapelle-aux-Saints Neanderthal, Piltdown Man, and modern *Homo sapiens* in *A Guide to the Fossil Remains of Man*, published by the Department of Geology, British Museum (Natural History), 1918.

Woodward's overarching theories. (Casts of Piltdown—used even today as teaching materials or historical curiosities—are based on Smith Woodward's reconstructions.) Interestingly, the cast of Piltdown caused some issues within scientific communities, as not all researchers were satisfied with examining a cast of a fossil and not the fossil itself. In his 1915 assessment of the Piltdown remains, Smithsonian scientist Gerrit Smith Miller, Jr., complained about having to use a cast of the remains for his study. Even with the cast, Miller concluded that the skull fragments and jaw were simply too different from each other to assume that they were from the same individual. Miller posited that the cranium was from a human and assigned the jaw to the proposed species *Pan vertus*, a species of fossil chimpanzee suggested by Miller.

But most of Piltdown's public met the fossil through art and museum exhibits, not necessarily through the fossil's cast replicas. Sketches of Piltdown weren't hard to come by—every newspaper article that mentioned the fossil seemed to have some kind of artistic doodle of Piltdown's face to put with the article. But it was Belgian museum conservator Aimé Rutot's reconstruction of the fossil that came to dominate the paleoart genres and museum scenes of the early twentieth century. The reconstruction was part of a series of

This stereoscope reconstruction of Piltdown by Keystone would have been a way for people to "see" the Piltdown exhibit.

sculpted busts of prehistoric humans, produced in Belgium in the 1910s, that were widely disseminated throughout the 1920s (either as copies or in photographs), and it was this reconstruction that became Piltdown's most public face.[14]

Rutot's reconstruction pushed public awareness of the fossil even further when the Keystone View Company included Piltdown as one of the stereoscope cards in its biology unit as a teaching tool. This specific card—"Evolution, Early Man: Piltdown"—puts the Piltdown fossil squarely in the public's eye on two levels. Not only does Rutot's reconstruction put a face on the fossil, the Keystone stereoscope cards reinforced the accessible nature of the image—no scientific expertise was needed to use the instrument or to interpret the image. The Piltdown specimen could be studied, photographed, and sketched and the cast propagated through scientific, educational, and museum circles, lending a sense of credibility—even legitimacy—to the fossil copy.

———

Although Piltdown appeared to offer perfect evidence for the British-driven and British-centered view of human evolution, several aspects of the find were troubling to many within scientific communities. Some were concerned that the bones were conveniently missing their most diagnostic features, while others were concerned about whether the gravels that the fossil was found in were really as ancient as the Pleistocene.

Scientific responses to the Piltdown fossil varied, even in that initial meeting in December 1912. Two major issues about the fossil were immediately raised by the discussants, including the prominent anatomists Arthur Keith and Grafton Elliot Smith and archaeologist William Boyd Dawkins. First, they were concerned about the association of the skull with the jaw—whether the recovered and fragmentary parts belonged to the same species, let alone the same individual. Second, opinion was split regarding the age of the fossil—whether the Piltdown discovery was rather recent in age, from the Holocene, or older, from

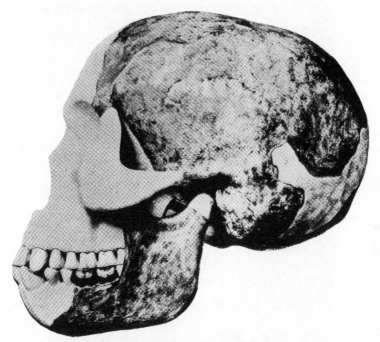

Skull of *Eoanthropus dawsoni*, or Piltdown Man. The smooth white sections are reconstructed parts of the cast, while the darker brown sections are replicas of the Piltdown remains. *(Wellcome Library, London)*

the Pleistocene. If the gravels and surrounding materials could be undisputedly sourced to an older geological epoch (say, the Pliocene or Pleistocene), then logically the fossil materials that were recovered from those sediments would geologically be associated with the older materials—indicating that Piltdown Man was a legitimate fossil, old enough to be a contender in the hominin family tree. Indeed, as British Museum anatomist Grafton Elliot Smith noted, why would one assume "that Nature had played the amazing trick of depositing in the same bed of gravel the brain-case (*without* the jaw) of a hitherto unknown type of early Pleistocene Man displaying unique, simian traits alongside the jaw (*without* the brain-case) of an equally unknown Pleistocene Ape displaying human traits unknown in any Ape"?[15]

The Piltdown man of Sussex, England. Reconstruction of bust at ¾ view, sculpture by J. McGregor, 1927. *(Wellcome Library, London)*

In his popular book *Missing Links*, John Reader notes, "The experts may have disputed the association of the Piltdown jaw and skull . . . but the Piltdown remains proved beyond doubt that mankind had already developed a remarkably large brain by the beginning of the Pleistocene. And the implications of this were very important."[16] Most important to Piltdown's success was that these experts elevated the evolutionary significance of the Piltdown fossil above the *Pithecanthropus* fossil in Java and the La Chapelle Neanderthal—because of Piltdown's larger brain and its clean geological context. Piltdown's solid footing in British scientific circles made the fossil particularly difficult to challenge until new fossils were discovered in Zhoukoudian in China (described by Franz Weidenreich in the late 1930s), providing paleoanthropology with a more complicated evolutionary tree.

By the late 1940s, the rumblings of discontent about Piltdown within academic circles deepened. Archaeologist Alvan T. Marston, for example, gave a paper at the Geological Society of London in 1947, where he described the Piltdown mandible and canine tooth as "pure ape"—a claim that, if true, would mean that the fossil wasn't a human ancestor. (Marston had discovered a Pleistocene hominin cranium at Swanscombe—a site in Kent, England—in the mid-1930s; his participation in scientific meetings as an elevated amateur would not be as odd as it might seem to modern readers.) Marston's claim prompted a great deal of discussion and added to earlier concerns, like those expressed by Gerrit Smith Miller, Jr., of the Smithsonian. Dr. Kenneth Oakley, a geologist and paleontologist from the British Museum (Natural History), suggested that it might be possible to test the Piltdown fossils for fluorine content using a method he himself had developed, and this test would help to resolve the community's questions.

Testing the Piltdown fossil—any fossil, really, but Piltdown specifically—using Oakley's criteria meant comparing the fluorine content of modern, subfossil, and fossil materials of specific ages. Fluorine testing does not present researchers with an absolute date, as would be the case with carbon-14 testing or other radiometric testing, but it does indicate whether the materials being tested are the same age. If tested materials showed the same amount of fluorine, then the materials would be the same age, since the materials would have absorbed the same amount of fluorine from their environment. This logic had been applied to the femur, skullcap, and tooth Dutch anatomist Eugène Dubois had recovered in the late nineteenth century in Java, indicating that the Java individual was just that: a single individual. Oakley's method of testing the Piltdown materials—the cranium, jaw, canine, and other mammalian fossils from the collection—meant researchers would be able to know whether the Piltdown fossil parts were really from one individual, as assumed, or not.

Fluorine testing requires a small part of a fossil be destroyed in order to measure the amount of fluorine in the specimen. In September 1948, after months of careful consideration, the British Museum's Department of Geology gave permission for Oakley and his associates to sample part of the Piltdown fossil for their analysis. "The curator of a palaeontological collection, which may contain rare specimens of great scientific importance, is frequently faced with the problem of whether to allow such specimens to be reinvestigated by treatment with acids, section, removal of fragments for chemical analysis, or other methods which might seem to involve damage to a unique object," notes W. N. Edwards, keeper of geology for the British Museum (Natural History), in the 1953 publication *The Solution of the Piltdown Problem.* "The cautious attitude of a previous generation has undoubtedly preserved for their successors many fossils which, for examples, might have been damaged by mechanical treatment in the past, but can now be developed in perfection by more recently devised chemical methods."[17] In *The Piltdown Inquest*, author Charles Blinderman describes the sampling: "This wasn't as much of a desecration as drilling into the Crown Jewels, but the fossils had been protected from German bombs during two wars, from being molested by inquisitive scientists for forty years, and even from the public, who viewed not the fossils themselves, but casts."[18]

The first round of fluorine tests indicated that the Piltdown materials were of a similar age and differed from the elephant and hippo fossils excavated from Piltdown. But the results also showed a difference in fluorine content between the crania and mandibular fragments. Subsequent chemical analyses measured nitrogen in the Piltdown materials and indicated that the pieces were much too recent to have come from the Pleistocene. The Piltdown "fossil" was made up of bones from three modern species—a human skull, an orangutan jaw, and chimpanzee teeth. Under the lens of powerful microscopes, the teeth in the mandible showed striations across their surfaces—evidence that the cusps on the ape molars had been filed down to make a correct identification of their species difficult. And thanks to this new scrutiny of the "fossil," researchers found that the entire set of bones was stained with a dark iron solution to make it look older than it was. The findings were conclusive: Piltdown Man was a fake.

"From the evidence which we have obtained, it is now clear that the distinguished palaeontologists and archaeologists who took part in the excavations at Piltdown were the victims of a most elaborate and carefully prepared hoax," anthropologists Kenneth Oakley, Joseph Weiner, and Wilfrid Le Gros Clark argue in the report of their findings, *The Solution of the Piltdown Problem*. "Let it be said, however, in exoneration of those who have assumed the Piltdown fragments to belong to a single individual, or who, having examined the original specimens, either regarded the mandible and canine as those of a fossil ape or else assumed (tacitly or explicitly) that the problem was not capable of solution on the available evidence, that the faking of the mandible and canine is so extraordinarily skillful, and the perpetration of the hoax appears to have been so entirely unscrupulous and inexplicable, as to find no parallel in the history of palaeontological discovery."[19]

Undeniably, learning that the Piltdown fossil was a hoax stirred up a lot of emotion among scientists who had worked with the fossils, particularly since Piltdown Man had been so firmly anchored in the evolutionary family tree for so long. Although Oakley and his colleagues had uncovered the truth about Piltdown, there was a respect for the previous generation of scientists who had participated in the early days of the research. One of the more poignant moments of *Eoanthropus*'s fall from grace came when Oakley, his wife, and a few others from the museum went to tell Sir Arthur Keith—himself an ardent supporter of Piltdown as an evolutionary ancestor—of their findings. Keith's correspondence from that time paints the picture of a frail old man with hesitant, shaky handwriting—long retired from the museum, but still curious about the world and his beloved fossils. It was almost as if they were going to break the news that a colleague had died. And, in some ways, one had. Keith had lived and thought about Piltdown for forty years. In a moment of quiet gravitas, Keith remarked how grateful he was that Sir Arthur Smith Woodward wasn't alive to find out that Piltdown was a fake.

"It isn't hard to see why Smith Woodward—and so many others—was taken in so overwhelmingly. Given the desire to find evidence of ancient man in Britain, why should he even question the veracity of the specimens he himself had seen picked up from the gravel bed?" historian of science Dr. Karolyn Schindler notes, brilliantly contextualizing the discovery. "The question of course remains as to whether it was Smith Woodward's great eminence that lent too much credence to the discovery, yet the majority though not all of the distinguished scientists involved in Piltdown had no doubt of its antiquity. Who, after all, would suspect a hoax like this?"[20]

Once the fossil was unveiled as a hoax, the question that burned in everyone's mind was: *Who?* Who had perpetrated this elaborate hoax?

The list of suspects was—and still is—long. Many considered the prime possibility to be the fossil's discoverer, Charles Dawson. Others have proposed prominent scientists William J. Sollas and Sir Arthur Keith. Still others suspect the archaeologist-philosopher Pierre Teilhard de Chardin. Even the celebrated Sir Arthur Conan Doyle has his share of claimants suggesting that he had perpetrated the hoax, since he visited the site several times. For the most part, however, suspicion has centered around Charles Dawson. Dawson, the argument goes, must have been behind the hoax—desperate for the scientific recognition and celebrity that was attached to a famous fossil. Over the decades, though, support for Dawson's character came from a variety of sources. An indignant letter to the editor of *The Times*, dated November 25, 1953, came from F. J. M. Postlethwaite, Dawson's friend and stepson. He spoke on behalf of the late Dawson, gave a character witness, recalled watching Dawson's excavations in 1911 and 1912 while on military leave from the Sudan—everything was on the up-and-up— and claimed Dawson could never be tied up in such an unseemly fraud. "Charles Dawson was at all times far too honest and faithful to his research to have been accessory to any faking whatsoever. He was himself duped, and from statements appearing in the Press, such is evidently the opinion of those who knew him well, some of whom are scientists of repute."[21]

———

At the point that the fossil was exposed, both scientific and popular audiences needed to find some way to make sense of such an elaborate hoax that had fooled so many people for so long. Some scientists—including Weiner and Oakley who debunked the fossil— started writing op-ed articles and piecing together oral histories and interviews with people involved with the entire Piltdown scenario. Kenneth Oakley, perhaps more than any other researcher associated with Piltdown, took it upon himself to collect these histories in an attempt to solve Piltdown's mystery. In an interview with one of Dawson's legal lackeys who ran Dawson's office, the gentleman recalled the difficulties of working with an amateur naturalist during the Piltdown field seasons decades before: "On occasion Mr. Charles Dawson boiled specimens in the office kettle. On these days I had to delay making the office tea."[22]

However interesting the fossil was as a human ancestor, it was that much more salacious and exciting as a fraud. The social pull of

Piltdown reached so far beyond any kind of fluorine or chemical tests—as such, "Piltdown" has impacted people for decades, long after the hoax was revealed, even people without any kind of scientific or expert stake in the scientific evaluation of the specimen. In popular vernacular, "Piltdown" has come to be synonymous with an elaborate "fraud" or "hoax."

The entire public life of Piltdown can be found in the bits and pieces of everyday effluvia—the outraged letter to the editor protesting the besmirching of a friend's or colleague's reputation; the odd bit of historical satire and poetry; the cartoon that make fun of the fossil and its adoring public. One of the folders in the British Museum's Piltdown Collection is marked "Humour" and contains a trove of documents and sketches that speak to the lighter side of the fossil and its story. In 1954, Mr. N. P. Morris, a colleague of Kenneth Oakley's, told the entire Piltdown story through campy prose:

> Some forty years ago the Piltdown bones
> Were found among some gravel, sticks and stones,
> And when assembled—after quite a lull—
> The Press announced the famous Piltdown Skull.
> So *E-o-anthropus* achieved his fame
> (Together with an academic name)
> For then it was the experts dared to think
> That they had really found the Missing Link.
> The world in general got a nasty shock
> On hearing thus of their ancestral stock,
> But, by and large, they settled down again,
> Their hopes of nobler vintage down the drain.
> But as the years rolled by the specialists—
> The Archaeo- and Anthropologists—
> Grew more and more convinced they'd been beguiled,
> And "Eo" was, in fact a Problem Child.

The mandible, according to their test,
Was modern ape, not fossil like the rest,
Which proved, undoubtedly, the knotty point,
That Piltdown's jaw was sorely out of joint,
The battle rages as hotly as before,
For Piltdown (pictured in the Press once more
With lower dentures not as they should be)
No longer held the evolution key.
The scientists, still far from satisfied,
Went down to Sussex where the victim "died."
From water and the gravel round about
They found his nationality in doubt.
An African? a D.P.—it would seem?
His politics! who knows what they'd have been!
While all these years this cur of wide renown
Had claimed protection from the British Crown.
(For radioactive tests have now revealed.
Some facts on vintage, hitherto concealed,
And X-ray analysis has shown
What clever chaps could do to fossil bone.)
So now with customary British phlegm
We chant for him a final requiem,
But who laughs last, may laugh the longest peel—
The hoaxer's ghost has not been brought to heel.[23]

The poem highlights a lighter side to the Piltdown saga and helpfully gives us all the significant parts that made Piltdown, well, Piltdown. An academic name. Ape. Problem Child. Missing Link. Fame. Clever chaps. Sussex. Radioactive tests. British phlegm. Humor, wit, and satire have as much a part in the Piltdown story as the fossil's fluorine tests and museum displays. As the public came to embrace a fraud, the Piltdown story coalesced and the fossil's identity began to change.

The question of how to respond to the hoax fronted the British political establishment as well as its scientific communities. Only days after the news of the hoax broke, there was actually a motion before the House of Commons that concerned Piltdown. One member of the House had taken up the cause "on behalf of decades of disappointed schoolboys." The motion read, "That this House has no confidence in the trustees of the British Museum, other than the Speaker of the House of Commons, because of the tardiness of their discovery that the skull of the Piltdown Man is a partial fake." The motion never had a chance of passing, and the Speaker, with poorly suppressed hilarity, offered the observation that "the honorable trustees have many other things to do besides examine the authenticity of a lot of old bones."[24]

The British Museum (Natural History) found itself thrust into the spotlight not only to field inquires from House trustees on behalf of disappointed schoolboys, but also to deal with the Nature Conservancy, which, in 1953, had just allocated conservancy status, funds, and legitimacy to the Piltdown site, designating it as a nationally important site for Britain's archaeological heritage. The status was quickly and quietly withdrawn: "Piltdown Site Is Handed Back," read a headline in the *Evening News* on November 24, 1954. (The site was, however, formally gifted to the Nature Conservancy in April 1957.)

The museum even had to figure out how to deal with the conspiracy wingnuts who came out of the woodwork, making general pests of themselves, offering their take on "solutions" to the Piltdown problem. A certain Mr. Alfred Scheuer, for example, created such a stink among the museum staff—with misspelled, poorly written, flight-of-fancy slanderous missives toward those associated with Piltdown, claiming that the museum had faked other finds—that the staff eventually stopped responding to Scheuer's crazed letters. A note in the museum's "Scheuer File" has a document dated April 28, 1967, from secretary Rosemary Powers: "Dr. Oakley: Mr. Jessup brought in this correspondence he had with Scheuer, so that we might squash the blighter with it if he pops up again. He has not been heard from in 3 years, happily. I have appended the old file number AL 1955/10."[25]

More than its altered status as an evolutionary ancestor, photographs and portraiture of Piltdown changed considerably over the course of the specimen's life. Once Oakley and his associates began working with Piltdown, the fossil was photographed surrounded by laboratory equipment. Scientists in formal wear weren't cradling the fossil; rather, scientists in white lab coats used instruments to interact with Piltdown. Piltdown was no longer "Piltdown Man," a human ancestor; it had become a specimen and scientific object to be poked, prodded, and worked out. When newspapers from the 1950s onward ran stories about Piltdown, they used photographs of the specimen in its laboratory contexts, and the public's views of Piltdown were shaped by that media lens.

———

So what's the proper place in public and scientific spaces for something like Piltdown? A museum? Perhaps. The specimen is certainly famous, and it's a significant part of paleoanthropology's history. But if a museum's job as an institution is to lend a certain credibility and legitimacy to the materials displayed in it, displaying Piltdown, or even a cast of Piltdown, becomes problematic, especially without proper context.

Kenneth Feder, author of *Frauds, Myths, and Mysteries: Science and Pseudoscience in Archaeology*, recalls a trip to the London Natural History Museum to check out Piltdown in its native museum habitat. "When I had trouble finding the fossil in a museum case, I approached a woman at the front desk, asking where I might see the Piltdown remains," he explains. "'Oh, that is not on display, sir,' and went on to inform me, rather condescendingly, 'It was all rubbish, you know.'" (Feder does note that Piltdown remains are brought out on display on occasion, particularly when relevant as archaeological fakes.)[26] This begs the question of what purpose these casts serve, beyond historical curiosities. The Sterkfontein Museum, in the Cradle of Humankind, South Africa, for example, has an interesting take on showing a cast of Piltdown to its audiences. The cast is displayed with the note, "Piltdown. Fake skull. Sussex."

"Piltdown Man", fake skull Sussex, England, 1912

While Piltdown is a fake fossil, some museums opt to display it as part of paleoanthropology's history. The exhibit in Sterkfontein, South Africa, allows visitors to see that Piltdown was a significant discovery. *(L. Pyne)*

It seems that just about everyone professionally associated with paleoanthropology today—and certainly every paleoenthusiast—has a theory on why the hoax lasted as long as it did (forty years) and what Piltdown meant (and means) to paleoanthropology. Consequently, talking about Piltdown feels a bit like asking someone to share her thoughts about a faked moon landing or conspiracies around JFK's death. Even my trip to the Natural History Museum archives, so I could examine the Piltdown files myself, netted a polite smile from the ever professional archival librarians. After they wheeled over three carts full of Piltdown materials, they offered the observation that those files are "always extremely popular." I felt almost as if I had just requested the secret files of the Illuminati.

But the drive to figure out who perpetrated the hoax and how the guilty party was so successful underscores why we are still talking about Piltdown today. In 2012, to mark the centenary of the fossil's discovery, a team of fifteen interdisciplinary scientists associated with the Natural History Museum in London—calling themselves Piltdowners—met with the intent of cracking opening the Piltdown mystery, treating the forgery rather like police investigating a cold case. The team of scientists—made up of paleoanthropologists, archaeologists, and paleontologists, as well as geneticists and museum curators—have taken a very twenty-first-century approach to unpacking the fossil forgery, treating the entire episode rather like the investigation of an art crime rather than the plot of an Agatha Christie novel. Where efforts in earlier decades had focused mainly on identifying the perpetrator of the hoax, twenty-first-century studies have worked to understand the context of the fakery. Simply asking who perpetrated the hoax doesn't begin to unpack the complexities of how it has lived within paleoanthropology's own history. "The Piltdown conundrum is addictive, totally addictive," archaeologist Dr. Simon Parfitt confessed in an interview with *Evolve* magazine.

At the Piltdowner meeting, Natural History Museum curator Dr. Rob Kruszynski offered a rundown of the impressive number of scientific tests that the material from the cold case had been subjected to, including nearly twenty different forms of analysis over sixty years. After the fluorine and radiocarbon testing exposed the hoax in 1953, the litany of analyses ramped up significantly. These new tests and methods—like confocal microscopy and CT scanning—offer new evidence about the Piltdown specimen to the ever growing body of Piltdown literature; the opportunity to showcase new and different analyses and methodologies on such a famous specimen undoubtedly holds a particular social cachet.

———

Like the Old Man, the Piltdown "fossil" has spawned an impressive amount of literature outside of strictly scientific publications. Ronald Millar's 1972 book *The Piltdown Men* presents the entire story in excruciating detail, with nothing too trivial to include. Other pillars of the Piltdown's canon include J. S. Weiner's 1955 *The Piltdown Forgery*; Charles Blinderman's 1986 *The Piltdown Inquest*; Frank Spencer's 1990 *Piltdown: A Scientific Forgery*; Miles Russell's 2003 *Piltdown Man: The Secret Life of Charles Dawson and the World's Greatest Archaeological Hoax*; John E. Walsh's 1996 tome *Unraveling Piltdown: The Scientific Fraud of the Century and Its Solution* . . . to say nothing of the countless chapters, pamphlets, articles, monographs, and blog posts that fill the archives and serve as fodder for conspiracy theorists. I don't think that there is a more published, more examined specimen in the fossil record, from dinosaurs to hominins, than Piltdown.

Plenty of fossils have held honored places in hominin phylogenies (key species, if you will), and have been removed as narratives of human history are revised and rewritten. What makes Piltdown so unique is the very fact that it is a hoax, and its salacious infamy has guaranteed it a historical identity. In *Myth and Meaning: Cracking the Code of Culture*, anthropologist Claude Levi-Strauss suggests that story—narrative—provides the cultural backdrop to making sense of history; in this case, scientific history, like Piltdown, is no different. "Mythology is static, we find the same mythical elements combined over and over again. . . . [History] shows us that by using the same material . . . one can nevertheless succeed in building up an original account for each of them."[27]

By mere circumstance of the fossil being a fraud, it generated another social strata of meaning, an interpretation outside of science.

There were questions about the legitimacy of the Piltdown memorial (How should people commemorate something that was fake?), what to do with museum material (Should something "wrong" be displayed in a museum?), and what the repercussions would be to those accused of being involved with the hoax (What is slander and what is speculation?).

We tend to get so caught up in the mystery and intrigue of the fabrication that we overlook other aspects of its life. Its life, however, is more than just the initial moment of its discovery, the controversies about its legitimacy, and its contributions to scientific debates. The Piltdown fossil tells us about the business of how paleoanthropology "does science" and how that changes over time, correcting for errors and introducing new technologies and methodologies. (Piltdown as a cautionary tale is well situated even in popular culture; in the first season of *Bones*, for example, one of the anthropologists is told, "It's like Piltdown," to connote a potential hoax.)

To pretend that the story of Piltdown is simply or only a scientific narrative that hinges on a correct interpretation of evidence—debunking a false hominin ancestor—misses the whole picture. It is not less famous for its hoaxy status; rather, its celebrity seems to grow the more the fossil is studied and the less we know for certain about it. When the Piltdown committee published its report stating that the fossil was a hoax, speculation immediately swirled as to who could have done it. Many of the scientists involved with the chemical and other tests, alongside countless armchair historian-detectives, have spent time researching and gathering evidence to try to point to the hoax's perpetrator. But treating the Piltdown specimen like this reduces it to a material object. Concentrating on unveiling the hoax's perpetrator basically reduces the fossil to a somewhat embarrassing historical deus ex machina in the story of paleoanthropology. It's as if the fossil has no identity or purpose outside of its status as a hoax. Casting accusations—however founded or unfounded—about the hoaxer's identity also speaks to the social life and place that Piltdown occupied. In short, it is part of its identity. Even the terminology of the fossil hinges on that historical turning point—it moves from being Piltdown Man, graced with a scientific name, *Eoanthropus dawsoni*, and imbued with scientific meaning, to simply being "the Piltdown Hoax." The "Piltdown" part of its name and identity endure—forever linking the specimen with its place of origin—but the status changes. It's not an ancestor; it's an object. It's not an active piece of hominin phylogeny; it's separated from evolutionary discourse by the distancing article "the" and the distaste and notoriety that swirl around its hoaxy status. Piltdown's notoriety has essentially made it the Milli Vanilli of the paleo world—a fake known for being fake.

It's easy to see how Piltdown's discoverer, Charles Dawson, and its champion at the British Museum (Natural History), Arthur Smith Woodward, were invested in the fossil, but the social and scientific investment in Piltdown extended far beyond the narrow confines of scientific literature. Piltdown made its way into museum exhibits, education materials, postcards, satirical cartoons, and letters to the editors of various newspapers. People—culture writ large, really—were and still are invested in the fossil in a way that extended its influence far beyond its discovery site or home in the British Museum. "Our fascination with where we came from is boundless," Karolyn Schindler argues. "That was why Piltdown was so clever and successful as a hoax: it was what everyone wanted to find—or at least, it appeared to be."[28] In addition to these concerns raised at the fossil's initial presentation in 1912, the scientific establishment did not accept the fossil or its interpretation completely at face value. Others, though, like Gerrit Smith Miller, Jr., at the Smithsonian, questioned the geological antiquity of the specimen as well as the integrity of the fossil's provenience.

Twenty-first-century Piltdowners—scientists, historians, enthusiasts, and amateurs alike—continue to grapple with Piltdown, working out finer and finer details of how the hoax was committed and searching for the conspiratorial holy grail that would point unequivocally to the perpetrator. "*Eoanthropus* is a name with no one to possess it," Miller noted after the fossil was debunked. While *Eoanthropus* might be a species empty of fossils, Piltdown is a specimen full of intrigue and possibility.

Piltdown's open-ended story is intriguing from a historical or even literary perspective—like a fade-out, it allows readers to embrace the implicit ambiguity and to puzzle out a solution for themselves. With so many loose ends, unresolved bits, and unsubstantiated rumors of its life—from its convoluted beginnings where Dawson's workers described finding a "cocoa-nut" to its modern *CSI*-like life as a museum curiosity—the story of this fossil is far from over.

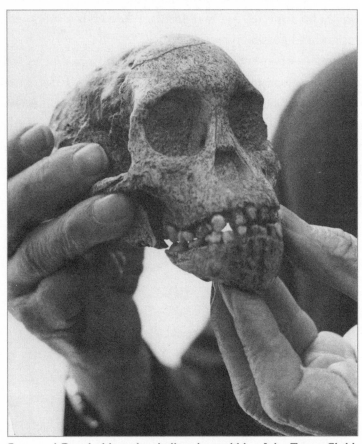

Raymond Dart holding the skull and mandible of the Taung Child.
(Raymond Dart Collection. Courtesy of the University of the Witwatersrand Archive)

CHAPTER THREE

THE TAUNG CHILD: THE RISE OF A FOLK HERO

A thrill of excitement shot through me. On the very top of the rock heap was what was undoubtedly an endocranial cast or mold of the interior of the skull. Had it been only the fossil-ised brain cast of any species of ape it would have ranked as a great discovery, for such a thing had never before been reported," Raymond Dart wrote in his memoir *Adventures with the Missing Link* in 1959, a little more than twenty-five years after discovering this remarkable fossil skull, the Taung Child, in 1924. "But I knew at a glance that what lay in my hands was no ordinary anthropoidal brain. Here I was certain was one of the most significant finds ever made in the history of anthropology. Darwin's largely discredited theory that man's early progenitors probably lived in Africa came back to me. Was I to be the instrument by which his 'missing link' was found?"[1]

At the beginning of the twentieth century, paleoanthropology's intellectual interests were firmly ensconced in Southeast Asia and Europe, thanks to the discovery of Java Man in 1891, several Neanderthals, like the Old Man of La Chapelle, and Piltdown Man in

England—in other words, just about anywhere but Africa. Dr. Raymond Dart, however, was in Johannesburg, South Africa, thousands of miles away from either hot spot in paleoanthropology. But Dart was right: the fossil he found was "one of the most significant finds ever made in the history of anthropology."

Today, the Taung Child is famous for its scientific significance as the first *Australopithecus africanus*, of course, but it is just as renowned for the ways it came to exemplify the intertwining of science, history, and the making of a paleocelebrity.

———

In January 1924, Dart was a young Australian anatomist beginning his career at the University of the Witwatersrand in Johannesburg, charged by the university to create a medical and anatomy department. Dart had spent two years prior studying neuroanatomy in London, under the mentorship of British neuroanatomist Sir Grafton Elliot Smith. At the end of his studies in London, made possible by a scholarship, the prominent anatomist Sir Arthur Keith persuaded Dart to apply for the newly vacant position in Johannesburg. Although Dart was rather horrified at the prospect of heading to South Africa, away from the scientific community in London, he successfully applied for the position with every intention of returning to London at some point in the future. (Keith would later write of Dart: "I was the one who recommended him for the post, but I did so, I am now free to confess, with a certain degree of trepidation. Of his knowledge, his power of intellect, and of imagination there could be no question; what rather frightened me was his flightiness, his scorn for accepted opinion, the unorthodoxy of his outlook.")[2]

When Dart arrived at the University of the Witwatersrand, he began to establish academic curricula as well as the school's medical program. One of his more popular classes had students out collecting fossils and comparing the specimens they found with the bones of other, extant species as a means of identifying their discoveries. Dart encouraged his students to collect fossil curiosities for the class, and soon fossil animals trickled into the classroom's laboratory. In early 1924, Dart's only female student, Josephine Salmons, saw a particularly curious fossil being used as a paperweight on the director's desk at the Buxton Limeworks quarry, where a friend of hers worked. (A slightly differing account places the fossil on display on the mantel of the family fireplace, where it piqued Salmons's interest.) She could tell that the fossil was some sort of primate and guessed that there was some deeper evolutionary significance to the fossil than that of a mere curio, so she asked the quarry director whether her mentor, Professor Raymond Dart, could take a look at it. Dart's assessment of the fossil was that it was a very old cercopithecoid, or an extinct species of baboon.

Finding the fossil primate was tremendously exciting for Dart and his students because it meant that other primates could be part of the South African fossil record. As an anatomist interested in the structure and evolution of the human brain, Dart was keen to collect more specimens that could shed light on the early evolution of primate brains. Dart asked Salmons to convey his very active interest in any fossils discovered in the Limeworks mine and even proposed offering a small financial reward to any worker there who procured interesting specimens. The director of the Northern Lime Company—Mr. A. E. Spiers, himself an amateur enthusiast and collector of fossil curios—readily agreed to stockpile fossils, although he declined Dart's offer for monetary compensation. Thus, the director of Buxton Limeworks, Mr. E. G. Izod, set about collecting the more interesting fossils found by the mine's workmen, which, thanks to the region's rich limestone geology, were plentiful.

Fossils from the Limeworks quarry were collected and shipped to Dart back in Johannesburg that fall. In October 1924, Dart received a crate of fossils from the mine the day he and his wife were to host a wedding, with Dart as the best man. Upon the arrival of the crate, Dart's wife, Dora, was less than impressed. In his autobiography, he—rather paternalistically—described Dora's reaction: "I suppose those are the fossils you've been expecting. Why on earth did they have to arrive today of all days? Now, Raymond, the guests will start arriving shortly and you can't go delving in all that rubble until the wedding's over and everybody has left. I know how important the fossils are to you, but please leave them until tomorrow."[3] Concerns about guests aside, Dart immediately started rummaging through the crated fossils, in full formal Edwardian attire. He came across a small, fossilized primate brain that stopped him cold. He was so enthralled with the discovery—"a thrill of excitement shot through me . . . I stood in the shade holding the [fossil] brain as greedily as any miser hugs his gold, my mind racing ahead"—that the wedding party had to more or less drag him down to the ceremony, where a rather put-out groom expectantly waited for Dart to perform his duties as best man. Dart recalled, "These pleasant daydreams were interrupted by the bridegroom himself tugging at my sleeve. 'My God, Ray,' he said, striving to keep the nervous urgency out of his voice. 'You've got to finish dressing immediately—or I'll have to find another best man. The bridal car should be here any moment.' Reluctantly, I replaced the rocks in the boxes, but I carried the endocranial cast and the stone from which it had come along with me and locked them away in my wardrobe."[4]

A slightly alternate version of the fossil's discovery was offered by one Dr. Young, a colleague of Dart's. In an interview with the *Johannesburg Star* in 1925, Young described how he'd arrived at the Taung quarry after a set of blastings and found the face of the "missing link" fossil exposed from the rock with the brain portion nearby—a perfect fit of the two fossils. Young claims he carefully packed up the find and, upon returning to Johannesburg, handed the fossil over to

Dart. Dr. Young's claims never met with much traction outside of his interview, although Dart, in his 1925 *Nature* publication of the fossil, does credit Professor Young and Miss Salmons for their assistance in recovering the fossil.[5]

In order to remove the fossils—the cranium and mandible—from the tough brecciated limestone, Dart pilfered several pairs of his wife's knitting needles and sharpened them to form tools to pick precisely at the rock around the fossil. For the next three months, Dart used every spare moment to patiently chip the matrix from the skull. Then, two days before Christmas, the face of a child emerged from the rock. Dart wrote: "I doubt if there was any parent prouder of his offspring than I was of my Taungs [sic] baby on that Christmas of 1924."[6] It was immediately christened the Taung Child—Raymond and Dora Dart's fossil scion.

————

Forty days after freeing the fossil from its limestone chrysalis, mid-January 1925, Dart sent an anatomical description, a series of photographs, and a manuscript about the fossil to the journal *Nature*, which was quick to publish his report. Dart described the fossil as "exhibit[ing] an extinct race of apes *intermediate between living anthropoids and man*."[7] Based on its anatomy, Dart described the fossil as the child of an "ape-like" ancestor to man, a small-brained hominin that could already walk bipedally, or on two legs. Dart named the species *Australopithecus africanus*—the southern ape of Africa. In his description, Dart pointed out the stark anatomical contrasts of the fossil to other apes, like gorillas and chimpanzees. Dart saw these differences—such as where the spinal column was positioned—as clear evidence to strengthen his interpretation of the fossil as a small-brained biped. In addition to the anatomical details of the teeth, the mandible, and the position of the vertebral column, Dart took his interpretations one step further and claimed that the fossil species was clear evidence that Africa was the "cradle of man-kind" (Darwin's own terminology) and that this fossil, this Taung Child, was excellent evidence of a "missing link" that neatly secured fossils within an explanatory schema—Dart saw no reason to aim low in his expectations of the fossil's import.

Upon its publication, the British scientific establishment back in London—Sir Arthur Keith, Sir Arthur Smith Woodward, and Dr. W. L. H. Duckworth—published a review of the fossil, also in *Nature*, and expressed a cautious, almost taciturn, interest in the fossil, but voiced no support of Dart's interpretation of the fossil as a human ancestor. They were convinced that the fossil was some sort of baboon, similar to the earlier fossil finds from the area. The only visual frame of reference they had for evaluating the fossil was the small photographs Dart included in his *Nature* article. What they wanted were measurements, casts, and detailed quantitative com-parisons. What they got was Dart's florid prose. "We must therefore conclude that it was only the *enhanced cerebral powers* possessed by this group which made their existence possible in this untoward environment [South African paleoenvironment]. . . . For the produc-tion of man, a different apprenticeship was needed *to sharpen the wits* and *quicken the higher manifestations of intellect.*"[8] Even Professor Elliot Grafton Smith, Dart's mentor and champion, expressed cau-tious curiosity about the Taung fossil at best. Sir Arthur Keith, on the other hand, was quite vocal in his dismissal of the Taung Child as an evolutionary ancestor. In short, the fossil simply didn't fit.

Dart's rather fantastical description of the fossil in *Nature* left little room for the rigors of methodology demanded within the scientific community. Moreover, the community's reticence to accept the fossil was due, in part, to the dominant evolutionary theory of the early twentieth century. According to theory then accepted and in vogue, fossil ancestry *ought* to be apelike, big-brained ancestors from South-east Asia or Europe. (The establishment scientists were all firm proponents of Piltdown Man as an ancestor; Piltdown's anatomy supported the current trends in evolutionary thinking, and it would be more than two decades before the fossil was debunked as a hoax.) Dart's Taung Child was "wrong"—it was geographically unexpected and none of the fossil's characteristics were in favor. But another part of the fossil's nonacceptance was due to the way that Dart "did" science. The way that the fossil was described, the nonconventional taxonomy (mixing Greek and Latin in the name), to say nothing of his rather rococo narrative—Dart was flying by the seat of his pants in how he communicated his findings to the scientific community, and his style put the establishment's collective knickers in a twist.

Raymond Dart posing for portrait with pipe, white lab coat, microscope, skull, and Taung Child fossil—all the trappings of science. (*Raymond Dart Collection. Courtesy of the University of the Witwatersrand Archive*)

Shortly after publishing the description of the Taung Child in *Nature*, Dart commissioned a cast of the fossil, which included its three components: the endocast of the brain, the mandible, and the craniofacial part of the skull. (Dart described the fossilized brain as "startling" with "its convolutions and furrows of the brain and the blood vessels of the skull . . . plainly visible.")[9] To create the fossil's casts, Dart contacted the London-based R. F. Damon & Co., a company well established within anthropological and paleontological circles. Prior to the Taung Child, R. F. Damon & Co. had created casts and busts of Piltdown as well as Eugène Dubois's 1891 Java Man discovery and countless fossil casts of all sorts of animals. (And a decade after the discovery of Taung, it would create the casts of the Peking Man fossils from Zhoukoudian.) The casts were created and slowly began to circulate through museums, scientific labs, and other spaces.

Casts of the fossil meant that, regardless of the interpretations surrounding its evolutionary status, the fossil was being seen by many audiences. (Dart held the copyright to the Taung Child and earned royalties on every cast made of it.) When Dart worked with R. F. Damon & Co. to set the prices for the Taung Child, the director, Mr. Barlow, begged Dart to reconsider his position about the exorbitant cost of the cast, arguing, "The prices you suggested would result in killing the demand and would create in my customers a feeling of resentment which I am not willing to incur."[10] Dart conceded to Barlow's suggestions of a lower price.

At the cost of £15, cast replicas of the Taung Child were commissioned from R. F. Damon and sent to other museums, including the American Museum of Natural History in the 1930s. (Fifteen pounds in 1925 equates to roughly £800, or $1,250, today—a significant chunk of change, but feasible within many museum budgets.) With a great deal of diplomatic tap dancing on the part of R. F. Damon, a set of Taung Child casts were even sent to the Moscow Museum in 1933. Dart kept up a correspondence with other paleoanthropologists (like the eminent Franz Weidenreich, then working at the Zhoukoudian site in China), offering to trade a copy of the Taung Child's cast for a copy of whatever work that was currently relevant, in the process building the comparative collection at the University of the Witwatersrand. Dart was inundated with numerous requests, from Australia to Botswana, for copies of the Taung fossil as museums sought to display the famous fossil to visitors and to keep it as a scientific resource for their scientists. Royalties from the casts continued to trickle back to Dart over the ensuing decades.

Dart was quick to propose that one of the casts be submitted to the British Empire Exhibition in late 1925, writing to the Exhibition Committee's chairman Captain Lane with an extensive proposal. The Exhibition showcased goods produced and manufactured in the colonies, and the connections between colonial raw materials and technology, such as the expansion of railroads across India. Over 1924 and 1925, the Exhibition attracted twenty-five million visitors.[11] It was a way for the British Empire to highlight, promote, and show off industry, technology, and science, and to establish commercial and industrial ties across the empire. It was certainly a timely opportunity to promote the fossil.

Because Dart's original publication of the Taung fossil in *Nature* was met with so much skepticism, Captain Lane was hesitant to accept the loan cast for the Exhibition. He worried he'd look particularly foolish if the Committee decided to display a fossil cast that the scientific community thought to be insignificant. Prominent anthropologist Sir Grafton Elliot Smith, however, championed the display of the fossil: "It is unusual for an investigator to issue casts of his material before his full report has been published. The South African authorities therefore have done a real service to science by exhibiting the casts at Wembley now."[12] Although Smith was cautious about interpreting the Taung Child as a human ancestor, he called Keith's rhetoric in *Nature* an "outburst" and argued that the Exhibition was fortunate to have the opportunity to display the cast.[13]

Once the Exhibition Committee was reassured by Smith that Dart wasn't some unhinged crank, the Committee was most excited to show the cast, praising Dart: "We have had a good deal of attention drawn to this exhibit by the newspaper reports and we are indeed grateful to you for having framed such a nice cast."[14] (Dart opted to send a cast of the fossil, rather than the original, to safeguard the Taung Child against the perils of travel.) In the year since the fossil's discovery and publication, newspapers in England, South Africa, and as far away as Tasmania had played up the scientific rivalries between Dart and London's scientific establishment and the question of the evolutionary legitimacy of the fossil, creating a huge public interest in actually seeing the Taung Child at the British Empire Exhibition.

Dart had a clear idea about how he thought the Taung Child should be put on display. Prior to sending off the cast and his materials to the Wembley Committee, Dart brainstormed how to organize information relevant to the fossil so that the viewer could easily follow along with the exhibit. On University of the Witwatersrand letterhead, Dart doodled potential options for showcase sizes, highlighting a four-foot table that would extend out toward the audience. He wanted to include various human and ape crania for a quick visual comparison. He also suggested that diagrams of the geographic provenience—the geologic strata—of the Taung region be shown as a backdrop.

In these sketches, Dart penned "Africa: Cradle of Humankind" on the right side of his sketch—an elegant allusion to the historical legitimacy of his interpretation of his fossil. The "Cradle of Humankind"— the area that Dart had referenced in his *Nature* article—connoted

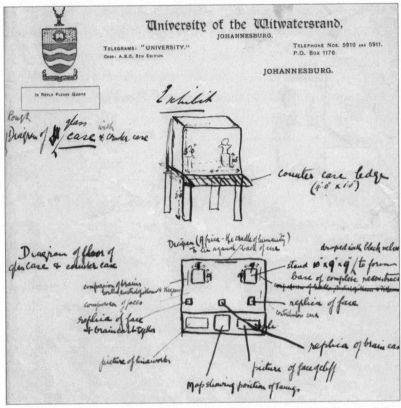

The Taung Child was displayed at Wembley as part of the British Empire Exhibition, 1925. These sketches show Dart's initial ideas about designing the public exhibit of the fossil. *(Raymond Dart Collection. Courtesy of the University of the Witwatersrand Archive)*

Darwin's theory of an African origin for humanity, rather than looking to Southeast Asia, the geographic locale very much in vogue within the scientific community thanks to finds like the Java Man. The reference to Africa, through Darwin's own language, neatly aligned himself and his discovery as Darwin-centric.

Dart's drawings included notes about where to hang photographs of the Limeworks cave and the cliff face to best acquaint viewers with an understanding of the geological context of the fossil itself. The collection of comparative crania curled around the left side of the case, and the entirety of the case's interior was to be draped in black velvet. The fossil cast was also a great equalizing object, as both expert and amateur fossil enthusiasts saw the fossil casts together in a space that did not differentiate the privilege of education and expertise. Sir Arthur Keith, much to his disgust, had to file through the Exhibition with the rather unwashed rank-and-file masses of

PROFESSOR DART'S EXHIBIT.

THE FOSSIL APE FOUND AT TAUNGS.

MAN'S NEAREST RELATION.

The fossilized skull of a hitherto unknown type of manlike ape, casts of which are now exhibited for the first time in this country, was blasted out of the limestone (50 feet below the surface and 200 feet from the original edge of the cliff) at Taungs in Bechuanaland in November, 1924, by workmen of the Northern Lime Company.

The discovery is exceptionally important and interesting. For the first time the whole face and form of the brain-case of a fossil man-like ape was revealed. Moreover, the Taungs ape was found in a place thousands of miles distant from the domain of the gorilla and chimpanzee, and in a region where forest conditions such as are essential to these other anthropoid apes seem to have been lacking. More important still, this ape, which like man may have been emancipated from the necessity of living in forests, seems to reveal definite evidence of nearer kinship with man's ancestors than any other ape presents.

Brochure for the Wembley exhibit describing the Taung Child, 1925. *(Raymond Dart Collection. Courtesy of the University of the Witwatersrand Archive)*

humanity for a quick glimpse at the fossil cast, and this did little to endear the Taung Child to him. His summary of the exhibit was less than flattering, and he doubled down on his claim that the Taung Child's species wasn't ancestral to modern humans: "A genealogist would make the identical mistake were he to claim a modern Sussex peasant as the ancestor of William the Conqueror."[15]

Meanwhile, members of the public—readers of the newspaper and amateur paleo and fossil enthusiasts—were curious about the fossil and queued for the opportunity to see it. Once the fossil and its scathing dismissal as "only" a fossil ape were published in *Nature*, newspapers from around the world rushed to publish the debate with synopses of the articles and the most current opinions about whether the Taung Child was, indeed, some sort of human ancestor. (One letter to an editor read, "Dear Sir, I wish for you to tell me whether the Taung Child is truly a human ancestor or not.")[16] These letters spoke to the desire of the general public to be able to classify—or at least make sense of—the fossil.

Although Dart received some brilliantly snippy antievolution letters from local South Africans concerned with the state of his immortal soul, overall the public adored the fossil and all it came to represent. Dart recognized how badly people wanted to know "the story" of the fossil they had read so much about in newspapers and seen at the Wembley exhibit. The act of publicly displaying the fossil—even a cast of the bones—folded the public into the fossil's interpretation. People became invested in the fossil and claimed a sense of agency about it.

———

By 1930, Dart accepted that he needed to work through a more conventional scientific process if he wanted the Taung Child to be accepted by the scientific community as a human ancestor. He prepared a lengthy monograph about the Taung Child with detailed anatomical measurements and comparisons. Dart boxed up the fossil to travel to London to meet with Sir Arthur Keith and other prominent anatomists. Dart intended to argue his case that Taung was indeed a human ancestor, and a fossil to be taken seriously.

Because boat travel to England with the real fossil was risky, Dart took out an insurance policy from Joseph Liddle Financial Insurance Agents of Johannesburg to cover the skull while it was in transit in May 1930. (Dart's fears were not unfounded; in 1919, crates of fossils from the Zhoukoudian site in China had been lost when the cargo ship they were traveling on sank while going around the Cape of Good Hope.) The Joseph Liddle policy, which covered the marine travel of the Taung Child to and from Europe as well as one year of travel within Europe, required Dart to personally accompany the fossil while insured.[17]

Once in London, the scientific establishment's reception to the Taung Child was cordial but decidedly cool. No one was overtly rude or plainly dismissive—but nor were they sold on Dart's insistence that human evolution would show a small-brained, bipedal hominin as an ancestor, even with his careful studies. Dart painted a rather sad picture of his trip: "This was no setting in which to vindicate claims once daring but now trite. . . . I stood in that austere and chilly room, my heart bounding with the hope that the expressions of polite attention on the four score faces before me might change to vivid interest as I spoke. I realized that my offering was an anti-climax."[18]

It was almost as if the moment that the fossil could have caught the scientific community's imagination had passed, and they were now interested in other specimens. The "next new thing" in the fossil community had been excavated in Zhoukoudian, China, and the British anatomists were interested in the significance of these China fossils from Peking, which were much more clearly humanlike in ancestry than the Taung Child. The Taung Child's fifteen minutes of fame were up, at least for the moment.

Dart returned to South Africa and more or less left the business of active fossil hunting to others, like Dr. Robert Broom, and devoted his own time to building up the Department of Anatomy at the University of the Witwatersrand medical school. He worked with ethnographic projects around South Africa and began to build a skeletal collection that would eventually become one of the most extensive in the world. He also served as a forensic expert in several court cases in Johannesburg. Although Dart did continue to study and write about australopithecine fossils, like the Taung Child—especially pushing his theories about the bloody and violent underpinnings of human evolution in later decades—his motivation and interests overall seemed to shift from fossils to medicine and anatomy.[19]

—————

But fossils continued to come out of the fossil-rich limestone quarries in the surrounding Transvaal region of South Africa, and plenty of other fossil enthusiasts (expert and amateur alike) stepped into the picture. Dr. Robert Broom, a Scottish national, was a paleontologist whose life in South Africa consisted of his medical practice and his work cataloging fossil lizards from the South Africa Karoo region. (The biologist J. B. S. Haldane once described Broom as a man of genius, fit to stand beside George Bernard Shaw, Beethoven, and Titian. Broom's own biographer, George Findlay, suggested that Broom was about as honest as a good poker player.) Dr. Broom's involvement with fossil human ancestors, in addition to his research on fossil lizards, began in 1925 with his congratulatory letter to Dart on the magnificent Taung find. Two weeks after Dart received Broom's letter, Broom himself arrived—unannounced—in Dart's laboratory. With *Hamlet*-worthy theatrical flair, Broom dropped to his knees in front of the fossil "in adoration of our ancestor."[20]

During that visit to Dart's lab in 1925, Broom and Dart discussed

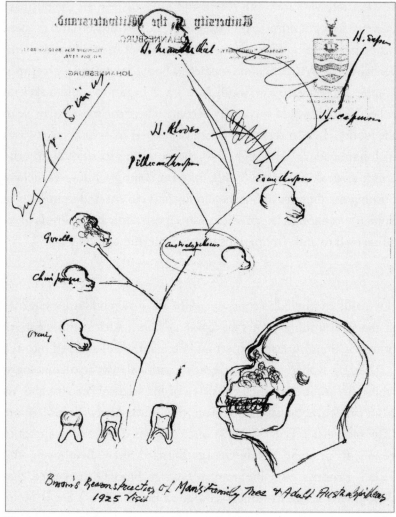

This hominin phylogeny doodle was sketched during a meeting between Robert Broom and Raymond Dart, 1925. *(Raymond Dart Collection. Courtesy of the University of the Witwatersrand Archive)*

different evolutionary scenarios for where Taung fit in the great schema of human evolution. Did Taung come before or after Piltdown? Was Dubois's fossil from Southeast Asia contemporaneous with Taung? No, it had a larger brain, so it had to come later? And what about the

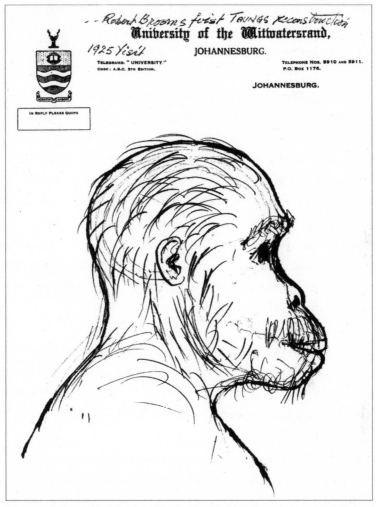

-- Robert Broom's first TAUNG reconstruction

1925 Visit

Raymond Dart's sketch of the Taung Child during a meeting with Robert Broom, 1925. *(Raymond Dart Collection. Courtesy of the University of the Witwatersrand Archive)*

Neanderthals? Where would they fall? While both Broom and Dart thought that the Taung Child was ancestral material, it was less clear how exactly it ought to fit in with other fossils. Broom's visit was more than just a chance to see the fossil; it emphasized the issues that would need to be overcome in order for the fossil to be accepted as a human ancestor.

One of the very real issues with the Taung Child was just that: it was a juvenile specimen, not fully grown when it died. As such, it was difficult to see how anatomical characteristics would be expressed as the species became an adult. In fact, Dart's use of the Taung Child as the type specimen of the species *Australopithecus africanus* provoked deep philosophical reflection on the nature of species and reconstructing species, even into the twenty-first century. (Since the fossil is a juvenile from the species, and not a fully formed adult, predicting how adults of the species would have looked made assigning adult specimens to the species difficult. If the type specimen—platonic ideal—was the Taung Child, then adult *Australopithecus africanus* individuals would be assigned based on how researchers thought an adult might have looked.) Broom realized this part of the problem: in order to really get at the anatomy and morphology of the fossil, one needed an adult specimen.

So Broom set off to find himself an adult australopithecine, and in 1947—more than twenty years after the original discovery of the Taung Child—he and his colleague John Robinson found one. The adult australopith they discovered in Sterkfontein was given the taxonomic name *Plesianthropus transvaalensis* ("Near-Man" of the Transvaal, nicknamed Mrs. Ples), and later reassigned to *Australopithecus africanus*. The change to *A. africanus* implicitly argued that the skull that Broom and Robinson had recovered belonged to the same species as the Taung Child. The fully grown adult specimen was finally accepted by the scientific establishment as being a valid species, and a species that might even show an ancestral relationship with *Homo*. Moreover, it answered the lingering questions from the paleo community about the problems of using a juvenile fossil to construct hominin lineages. Even Sir Arthur Keith had to admit, "You have found what I never thought could be found"—that is, a manlike jaw associated with an apelike skull, the exact reversal of Piltdown.[21]

In the late 1940s, back in Europe—Britain specifically—several major figures in the paleointelligentsia were taking issue with the decades-long interpretation of the Piltdown fossil. By this point, the Taung Child had been well casted, studied, and measured. Evidence that supported the fossil as a human ancestor was slowly mounting, including an incredibly favorable review of the fossil's anatomy in 1946 by Oxford anatomist Dr. Wilfrid Le Gros Clark. More fossils from a variety of geographic locales were complicating the phylogeny of human ancestry. Once Piltdown was completely debunked in 1953, it opened up intellectual space for Taung to occupy a place as a human ancestor.

With so much support for the Taung Child fossil, it was impossible to dismiss Dart's original interpretation of the fossil. "Professor Dart was right and I was wrong," Sir Arthur Keith conceded in the decades after the fossil's discovery and controversy. By 1985, the Taung Child—and *Australopithecus africanus*—were well accepted into the paleo pantheon as a legitimate hominin ancestor. Indeed, when the diamond jubilee of the Taung Child's discovery was held at the University of the Witwatersrand in 1985, Dart's reaction to all the fuss was a bit understated. The fossil had become folded into the mainstream of paleoanthropological practice. We have a plethora of ways to think about the life and death of an object, and this is no more apt than with the life of a fossil; in fact, there is something fantastically recursive about understanding the new life of something that itself was once alive. "What a wonderful occasion this is, isn't it? You know, I was never bitter about how I was treated back in 1925. I knew people wouldn't believe me. I wasn't in a hurry," said Dart at the jubilee celebration.[22]

———

The story of the Taung Child is practically apocryphal in paleoanthropology. These stories function as part of the science's own identity and values ("good science wins out over detractors"), but the stories also serve to create a heroic persona around Raymond Dart and the fossil itself. As the Taung Child moved into wider and wider audience circles from science to the public, it moved from a cast at Wembley into poetry, literature, parody, and just plain fun. Just as sagas and epic journeys are ways for audiences to become invested in the hero's quest, the journey of the Taung Child was embraced into a cultural narrative.

In the throes of Dart's arguments with the scientific establishment in the 1930s, one Dr. Walter Rose of Cape Town, a renowned herpetologist, composed a heroic saga of the fossil's story—entitled simply "Australopithecus."

> In Pliocene's far distant time,
> When good Earth was in her prime,
> In Africa's congenial clime
> I flourished.
> My Mother searched the earth for roots,
> From bushes bit the tender shoots,
> And I on these juicy fruits
> Was nourished.
> . . .
> The dust soon buried me from view,
> So I had nothing more to do
> But lie a million years or two
> Quite patient.
> And when for lime the earth was mined,
> They found me where I lay enshrined,
> And cried with joy, "My, here's a find
> Most ancient."
> Another cried, "'Tis plain to me
> This little creature that we see
> Is nothing but a chimpanzee,
> Believe me."
> "Tut, tut," retorted Dr. D . . .
> "My worthy colleague, have a heart;

You put the horse behind the cart.
You grieve me."

. . .

"I'm positive this little dome,
That in the forests used to roam,
Proves Africa was man's first home
Quite nicely.
I claim South Africa's the place
That first produced the human race.
This little skull confirms the case.
Precisely."
My finders sang triumphant songs
And said, "This cranium from Taungs
To the long-sought missing-link belongs,
We've found it
So we will claim the honor for
South Africa, whilst we explore
The neighborhood to find some more
Around it."[23]

And this, gentle reader, is a mere three stanzas from Rose's poetic gem. It touches on the significant elements of how the story of the Taung fossil became imbued into the public's mind. In Rose's enthusiastic telling, the fossil moves through the life and death in the flesh, its fantastical discovery, and its heroic battle in the scientific circles; and the significance of the fossil is projected into the future. It gives the fossil a significant frame of its celebrity—the fossil is a folk hero worthy of its own epic saga.

Between the lines, however, incredibly powerful elements move the Taung fossil sandwiched between scientific and popular spheres. There is an effort to root the fossil in a correct geologic era—"In Pliocene's far distant time." In Dart's copy of the poem, the word

"Pleistocene" is crossed out and replaced with "Pliocene." In another stanza, Dart himself is given lines through which he articulates his argument about Taung's place in the ancestral milieu. There is a subtle, almost nationalistic pride that Taung brings to South Africa. And most interestingly, the writing of the poem in the first person ultimately imbues the reader with a sense that the australopithecine possesses a sort of heroic agency. Taung chooses to overcome his environmental trials—he survives the harsh realities of the Pliocene paleoenvironment while his mother is eaten by a crocodile and dear old Dad is constricted by a python—to have his story told.

There is a striking humanistic flair in the setting of the poem. The first two stanzas set an idyllic and perfect Eden-esque backdrop for the australopithecine existence—no talk of segregation, no financial worries, no social ills that would be very much at the forefront of Rose's South African audience. After the environment stripped Taung and his older brother from their parents, Taung describes his death: "One day, while quarrelling for a bone / He [Taung's older brother] bashed my skull in with a stone / And left it in the cave, alone / To weather." This is an allusion to incredibly powerful literary motifs. One brother killing another—it's an archetype straight out of Genesis. In the space of one stanza, we see environmental determinism, an allegory of the Fall, the story of Cain and Abel, and the formation of Taung's lone heroic story—a fossil sage left to tell humanity about who they are and where they came from.

The explanatory power of a poem like this is immense, and its arguments and ideas demonstrate that there is more to the narrative of the fossil than what its morphometric measurements might indicate. The epic saga concludes—leaving us with the prose of heroic greatness.

Other types of stories about the fossil came to Dart, many of them completely unexpected. One of the more unusual was a very peculiar young adult novella, *The Fantasy of the Missing Link*, sent to him anonymously by a fan. The manuscript, which most likely dates from the mid-1930s, is signed "A Loyalist," its text championing Dart and the Taung Child. Set in Taung itself, the story opens with a miner, Ginger, grumbling about the fossils in the rock. "'Ere's another of them blinkin' old monkey fossils, turned up again, Joe—about the umpteenth this year, I reckon." He continues, "There must 'ave been a bloomin' regime of 'em 'ere some time or other, a regular monkey's harem."

A bit tangled and stilted, and more than a little confusing, *The Fantasy of the Missing Link* nevertheless gives us a very clear sense that the Taung Child was quickly working its way into the popular vernacular. The question of the fossil's place in the Great Chain—a human ancestor or just a "bloomin' monkey"—is crucial to the unfolding of the story. Joe Chambers, an educated miner, brings the fossil to the attention of Dr. Daye, who in turn comes out to the Buxton Limeworks Museum. Dr. Daye's and Joe's monologues in the story serve as asides for slipping in science to the audience, as the speeches are peppered with Darwin, evolution, and the nature of family trees.

Interestingly, *The Fantasy of the Missing Link* takes on evolution and Darwin as socially problematic, pitting evolution against religion, through Ginger, a gruff miner. Ginger is reticent to accept a Darwinian perspective of evolution: "So 'elp me Gaud, Joe, if you say again that I'm like that damned blasted old monkey fossil, I'll slash you acrost the face with this," he says, brandishing a pickax. "I'll make such a mess of your face, that yer own mother won't know you!" Ginger describes the Bible stories, like the Garden of Eden, as the only origin story he is interested in. (At the end of the book, when the recovered fossil is to be sent to England for further study, Dart crossed out the author's mention of Professor Elliot Smith and Sir Arthur Keith and penned in the names Professor Elland Swift and Sir Andrew Kelly—not, perhaps, the most subtle of aliases.)[24] Each part of the Taung Child's story—from discovery to location to Darwinian debates—is present within *The Fantasy*. The story finishes with the origin question—Darwin versus religion—juxtaposed and unresolved. *Inherit the Wind* the fan's novella is not, but its themes and sentiments certainly overlap.

———

In the 1940s, Dart began examining other bone and artifact assemblages collected by a local South African schoolteacher, Wilfred Eitzman, from several sites—like Sterkfontein and Makapansgat—near Taung. Both of these sites were rich in fossilized cores from antelope horns and shaped stone tools, raising the questions of who created these tools and for what purpose. Dart studied both assemblages several times and concluded that the fossilized bones and stone tools from the sites were created by the Taung Child's species and that these australopithecines were "predatory ape-men" bludgeoning their way across the landscape. He called this complex of stone and bone technology the "Osteodontokeratic Culture" (ODK) and published numerous articles arguing complex sequences and timelines of particular tools. In ODK culture, Taung and his ape-men were the hunters—the dominators—of the landscape.

Where Dart had imagined a violent, bloodthirsty, bone-club-wielding set of human ancestors, others in the scientific community (such as Dr. Wilfrid Le Gros Clark) argued that Dart's ODK culture pushed the limits of scientific evidence and interpretation. Le Gros Clark, himself a supporter of the Taung Child as an ancestor, argued that Dart's ODK depended primarily on a lack of alternative hypotheses for the scientific community to evaluate. (In other words, what would account for the accumulation of bones if not for the Taung Child's hominin species?) What Dart's hypothesis did, however, was help to usher in new fields of study within archaeology and paleoanthropology—fields of study, like taphonomy, that looked at how soils and bones and rocks accumulated in caves such as Makapansgat. These new studies, by researchers like Dr. Sherwood Washborn and Dr. Charles Brain, determined that natural causes accounted for the bones' accumulation. Brain's studies took this budding field of taphonomy one step further away from ODK-like interpretations by matching leopard teeth to puncture marks in a recovered australopithecine skull from another South African fossil site, Swartkrans. These tooth punctures, along with other findings, illustrated that hominins were vulnerable in the landscape— the hunter was now interpreted as the hunted.

In the public imagination, however, the idea of a savage human ancestry caught fire, thanks to Robert Ardrey's *African Genesis*, published in 1961, which argued humans were descendants of a bloodthirsty, weapon-welding predatory ancestor. *African Genesis* contained several direct references to Dart and his writing, and Ardrey posited that aggression—as would be expected in ODK culture—was the best model for understanding the barbarity of a killer-ape human ancestry. Science fiction author Arthur C. Clarke's "The Sentinel," written in 1948, a year after Dart's first publication on ODK, served as the basis for the femur-wielding furry ancestors in Stanley Kubrick's *2001: A Space Odyssey*. The interpretation of the Taung Child's species became imbued with meaning and morality. These themes became deeply entrenched in the public's mind and were long associated with fossils like Taung, however much science pointed to a hominin very much at the mercy of its environment.[25]

In addition to the cultural milieu that swept around the fossil, the public also came to know the Taung Child through the fossil's museum life as dioramas present viewers with a story about extinct species. (Recall that the poses, faces, and arrangement of Neanderthals in the 1930s dioramas in the Field Museum ensured that viewers came away with a story that Neanderthals were primitive savages.) A reconstruction of a fossil provides a visual dimensionality of muscle, skin, hair, and movement that imbues a sense of "realness" to a fossil that a mere description, however detailed, simply cannot match.

One of the most interesting dioramas of the Taung Child was constructed in the Ditsong Museum in Pretoria, South Africa. The dioramas, built in the late 1960s, provided visitors for fifty years with explicit and implicit stories of human evolution building on South Africa's immense australopithecine fossil records. Some of the dioramas were small scenes with toy-sized hominins, while other dioramas showed life-sized scenes as museum visitors wandered through the South African environment of three million years ago. (In 2013 the dioramas were closed to the public for cleaning, restoration, and reworking.) How we think about the australopithecines and how the fossil species interacted with their environment has changed a great deal since the dioramas were first built. If the dioramas are opened to the public again, these changing interpretations of the fossil record—the hunter? the hunted?—ought to be reflected in the stories they tell museum visitors.[26]

The same paleoanthropology field school that introduced me to the Taung Child introduced me to these fantastical dioramas at the Ditsong, then called the Transvaal Museum. My favorite diorama was one on the second floor, where a stuffed leopard dragged an adult australopithecine off to its lair, with the australopithecine skull lodged firmly in its mouth—blood, rather macabrely, dripping out of the cranial tooth punctures. The entire scene was completely and fantastically over-the-top. In another corner, a leopard perched in a tree and chewed on a juvenile australopith with hominin body limbs accumulating below the branches. A different section of the room showcased a set of four australopithecines with the story of a nuclear family: Mom and Dad playing with their kids, while keeping a watchful eye on the predatory raptors perched above them. A small, furry moppet, labeled "Taung Child," toddled after his other family members. Other scenes highlighted early tool use, as adult hominins brandish clubs. And a small in-wall diorama showed a young adult australopithecine stretching, greeting the morning, as others begin to awaken against a sun-kissed African horizon.

The Taung Child

The "Taung Child" (*Australopithecus africanus*) looks apprehensively at the eagle. It is thought that an eagle killed the Taung child, and other small primates, about 2 million years ago.

Reconstruction of the Taung Child. Ditsong Museum, Johannesburg, 2013. *(Justin Adams)*

These dioramas tell several stories—early hominins were no match for their environment, an easy target in the South African paleoenvironment. The dioramas fold in then current scientific research, as the leopard-with-skull exhibit speaks directly to Dr. Brain's work with cave taphonomy. The australopithecine nuclear family has immediate appeal, ascribing humanlike traits to this scene, as the viewer imagines adults protecting and playing with their offspring. These stories help allow for agency within the australopithecine community. In the broader context of human evolution, they make the australopithecines more like us, modern humans. They create sympathy and empathy with the species, because we recognize ourselves in these familiar scenes. Just as the brandishing club is a clear cultural motif, courtesy of Stanley Kubrick, dioramas allow human ancestors narrative space outside of a strict scientific circumstance.

Putting a face on the Taung Child began considerably early in the fossil's life. During Robert Broom's 1925 visit, Broom sketched an aged Taung Child in profile. He gave the adult Taung Child a very thick set of eyebrows, full apelike facial features, chimplike tufts of hair. He also gave it a bit of a quizzical expression. Dart kept another framed pen-and-ink sketch of the Child in his office at the university, one that showed it as a young, impish, Puck-like sprite of a hominid with a toothy grin.

Artistic representations of the fossil—whether through cartoon sketches or museum dioramas—take a static object and give it a face and body, thus allowing for more "understanding" through its art than if we were simply to read a descriptive placard about the fossil. Though the Ditsong scenes are only one example, the afterlife of the Taung Child is expressed through dioramas in a multitude of museums in a plethora of contexts—some better, some worse, but all tell audiences some story about the fossil.[27]

Where portraiture and photographs of the Taung Child offer an official, even formal, lens for the intersection of art and science, other artistic media became a way for audiences to meet the Taung Child fossil. *Life of Bone* was such an exhibit. Shown in Johannesburg at the Origins Centre in May 2011, it was an incredible success. The exhibit (and its accompanying book) highlighted the juxtaposition of art and science as envisioned by three South African artists—Joni

Brenner, Gerhard Marx, and Karel Nel. As the artists describe it, their work draws directly and indirectly from human and fossil bones and shows "how bones intersect issues of human origin, evolution, deep time, lineage, ancestry, and belonging."[28] Their work also draws heavily on South Africa's history.

Brenner's work with watercolors for the exhibit featured paintings of the Taung Child from all angles in a variety of dark, muted reds and blacks. In some pieces, drops of paint ran across part of the Taung skull, lending Brenner's unique voice to the Taung Child story. "Conversations, which often took place in the presence of skeletal remains and the casts of hominin fossils, reflected on ways of knowing, mapping and telling; on things we can and cannot know about our histories; and on the natural and social forces that have an impact on how we understand these material remains and ourselves," Brenner explains.[29]

In 2009, the University of the Witwatersrand Philip V. Tobias Fossil Primate and Hominid Laboratory inducted a curious artifact into its fossil vault as part of the Taung Child legacy. In 1925, after the British Empire Exhibition at Wembley, Dart had a small wooden box made to store the Taung Child fossil. Stained a medium dark brown, the only artistic details on the box are the delicate floral tendrils on the brass latch. The nicks and scratches on the outside speak to years of handling and transport. The box housed the three parts of the Taung Child fossil: the bony part of the face, a jaw, and the endocast of the brain. In August 1931, during Dart's visit to London, his wife, Dora, accidently left the fossil—allegedly in the wooden box—in a taxi in London. Dart relished telling the story of the shocked taxi driver opening the box, finding the fossil skull inside, and promptly turning it over to the London police. The police, in their own stunned turn, reunited the fossil with Dora the following morning.

In more recent decades, showing the fossil with the box was simply de rigueur as the box and the fossil have come to share so much history. When Professor Philip Tobias, himself one of Dart's students, showed the fossil to different groups, pulling it out of its box was part of the experience of seeing the Taung Child. Dr. Kristi Lewton, a physical anthropologist, recalled seeing the box as part of Dr. Tobias's demo. "I was struck by the juxtaposition of the figurative place of the Taung Child as one of the most important fossil discoveries in paleo-anthropology, with its literal place—stored in a modest wooden box in a locked vault that was essentially a closet. At the time I thought, 'Who knew this incredible discovery just sits in a closet?!'"[30]

After decades in the wooden box, the fossil was given a new acrylic box. When this was announced, media in Johannesburg turned up at the laboratory to witness the event. The retired box was given a specimen number in the Hominid Vault corresponding to Taung 1, the specimen number for the Taung Child itself—thus inexorably tying the box to the specimen. The box now sits, neatly labeled, next to the Taung Child.

The original storage box for the Taung Child. No longer in use, the box now has an accession and catalog number and is stored in the Hominid Vault at the University of the Witwatersrand, next to the Taung fossil itself. *(L. Pyne)*

The old wooden box entered the fossil archive, truly becoming a cultural extension of a fossil hominin discovered almost ninety years prior. Interestingly, it is the only "cultural" artifact in the fossil archive and has become an artifact turned relic. The rock breccia matrix that Dart pried the fossil out of with his wife's knitting needles is archived along with the fossil, and that box, which all currently reside at the Hominid Vault. Adding the Taung Child's box to the hominid laboratory says a lot about what is archived, how, and why—a testament to the Taung Child's own cultural history. As the laboratory stores many of South Africa's famous hominin specimens, the addition of the box poses an interesting juxtaposition of a scientific object and a cultural one and shows the fluidity of scientific collections. The hominid laboratory holds fossils—the physical, tangible fossils—but also holds the history, stories, and associations of those fossils with paleoanthropology writ large.

———

A fossil hero needs an audience, and the Taung Child certainly has many audiences. The current curator of the University of the Witwatersrand fossils, Dr. Bernhard Zipfel, has described the experience of observing people interact with the Taung Child fossil. "As curator of fossils, I am privileged to be one of the very few people who regularly have the opportunity to see and handle the Taung child's skull. When I show the skull to both scientists and non-scientists, the almost predictable expressions of wonder are clearly not only brought about by the scientific significance of this, the type specimen of *Australopithecus africanus*, but also by the sheer beauty of the little skull."[31]

There is an interesting relationship between a fossil of the celebrity caliber of the Taung Child and the introduction of new methodologies for measurement. The celebrity fossil is a pillar of the paleo community—it's well understood, well studied, and well internalized. Because so many other methodologies have used that one fossil as a test case, testing a new methodology becomes that much more significant. For example, when digitizing fossils was first introduced into paleoanthropology as a way of capturing information about an object in three dimensions, the Taung Child was one of the first fossils to be digitized. The three-dimensional scan was published as an almost artistic portraiture in *National Geographic* in 1985, corresponding with the fossil's diamond jubilee. When CT scanning was introduced, the Taung Child, again, was one of the first to be scanned. A famous fossil—even one as examined as Taung—does more than rest on its historical laurels in the fossil archive. It still asks and answers scientific questions. Paleoanthropologist Dr. Lee Berger offered this simple observation: "The Taung Child is iconic."[32]

The fossil maintains that status as yet more types of tests are conducted on it. It's paleoanthropology's Matthew effect as the studied fossils get studied more and the less studied fossils less so. And just as the more studied fossils become more studied, the fame that surrounds them becomes more obvious, more emphasized, and more present. The discoverers and researchers associated with something scientifically famous became scientific celebrities themselves.

Delving into the celebrity folk hero qualities that surround the Taung Child is tricky. It's not enough to simply say, "This is a famous fossil and, by virtue of it being famous, it's a celebrity as demonstrated through its heroic science." Fame doesn't work by syllogism. How we think about the fossil today is shaped by its discovery, of course, but it's also shaped by its history, its meaning, and its mystique. Kristi Lewton recalled her experience seeing the Taung Child fossil in person: "Seeing the Taung Child in life grabbed me—it was history coming to life. When I saw the fossil, in the early 2000s, Professor Tobias was a central figure in paleoanthropology—a living legend, really. Every one of us at that fossil demo had heard the story of the origin of the Taung Child. So to see the fossil in person was incredible."[33]

The Taung Child continues to influence its variety of audiences. The fossil is a compelling object from the early days of paleoanthropology, but it also speaks to the historical means of "doing science" that were at play in the early twentieth century. While the evolutionary relationship between human and *Australopithecus africanus* was resolved by the mid-twentieth century, questions of how Taung and his species functioned on the South African paleolandscape 2.5 million to 5.3 million years ago continue to fascinate the scientific community and capture the public imagination.

The achievements that made the Taung Child famous were the decades of study, the fall of Piltdown, and, as we tell the story, the dedicated, dogged determinism of Dr. Raymond Dart—efforts that eventually vindicated the fossil as a hominin ancestor. In the discipline's historical canon, the fossil itself represents an underdog fighting for a place of recognition as an evolutionary ancestor to modern humans.

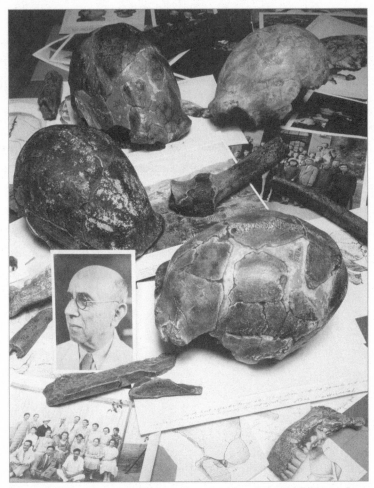

Composite image showing casts of hominid fossil skulls and bone fragments, drawings, and memorabilia brought by paleoanthropologist
Franz Weidenreich (center) from China to New York in 1941. These
fossil remains were recovered at Zhoukoudian between 1929 and 1937
and were classified by Weidenreich as *Homo erectus*, commonly referred
to as Peking Man. *(John Reader/Science Source)*

CHAPTER FOUR

PEKING MAN: A CURIOUS CASE OF PALEO-NOIR

In 2011, Dr. Per Ahlberg, Dr. Martin Kundrát, and curator Dr. Jan Ove Ebbestad began unpacking and cataloging the contents of forty boxes from fossil collections archived in the Museum of Evolution in Uppsala, Sweden. These boxes had not been opened since their materials had been packed off to Sweden from excavations at the well-known archaeological site of Zhoukoudian in China during the 1920s and 1930s. Among crates of the site's fossil fauna, the Swedish researchers found a hominin canine tooth. The tooth was chipped, the surface was very worn, and the dark brown root had broken off just below the gum line—but it was a tooth that looked surprisingly humanlike.

The Swedish scientists sent the tooth to their colleagues Liu Wu and Tong Haowen, paleontologists at the Institute of Vertebrate Paleontology and Paleoanthropology in Beijing, for analysis. Wu and Haowen determined that the tooth was a canine that would have belonged to Peking Man—a series of fossils excavated from Zhoukoudian in the first half of the twentieth century. Today Peking Man is taxonomically assigned to *Homo erectus*—an extinct Pleistocene

119

species in humans' evolutionary tree roughly 750,000 years old. But Wu and Haowen's description also meant that the tooth carried a certain historical distinction. Peking Man—as the assemblage of skulls, jaws, teeth, and other bones was collectively known—was one of the most celebrated fossils discovered at the beginning of the twentieth century. By identifying the tooth as part of Peking Man, the fossil tooth became a lost relic found.[1]

Recovered canine from Peking Man assemblage, Museum of Evolution, Uppsala University Archives, 2011. *(Museum of Evolution, Uppsala University, Sweden. Used with permission)*

It's hard to build a coherent narrative based strictly on disjointed details, and Peking Man's story is full of them: its fossils are a story built out of many stories, without a clear beginning, with many middles, and with no clear ending. Where other specimens like the Old Man or the Taung Child have a very specific moment of discovery and lives as scientific and cultural personae, Peking Man has only its many stories that are told and retold, forming a mythos of importance along nationalistic, scientific, and historical lines.

In the first decade of the twentieth century, paleoanthropology had very few fossils in its collections, and no fossils from mainland Asia. (The only Asian fossil in the historical record at that point was Eugène Dubois's discovery of Java Man, *Pithecanthropus erectus*, found on the Indonesian island of Java in 1891—*Homo erectus* to us today.) By the 1920s, the study of fossils and human evolution in China came from a variety of parties—China's budding interests in geology and anthropology, as well as investments from outside researchers interested in the artifacts and fossils that comprised China's archaeological and paleoanthropological records.

On the surface, Peking Man's story seems rather straightforward. In the summer of 1921, a young Austrian paleontologist, Otto Zdansky, found the first fossil hominin molar later classified as Peking Man while he surveyed the Zhoukoudian caves just forty kilometers outside of Beijing (then romanized as Peking). He picked up the molar and put it in his pocket. The fossil was eventually analyzed, along with other archaeological materials, and all skeletal materials were published in 1927's *Palaeontologica Sinica* as part of the new species *Sinanthropus pekinensis*, or Peking Man. On October 16, 1927, another *Sinanthropus* tooth was uncovered in the excavations, and Canadian paleoanthropologist Dr. Davidson Black felt confident that these fossils represented a completely new species of human ancestor. Over the course of a decade and a half, other *Sinanthropus* fossils like skulls, mandibles, teeth, and bone fragments were recovered from Zhoukoudian, enough fossils to represent a population of forty *Sinanthropus* individuals. Replica casts and museum displays were created, and national narratives written. Then, in December 1941, the fossils were lost during an attempt to ship them out of China before the invasion of the Japanese army. After the fossils went missing, they maintained their cachet through their casts and photographs. But the mystery of their disappearance and the question of where those original fossils went intrigued the scientific community and captured popular imagination, to say nothing of the interest of the government of China. All attempts to locate the fossils have ended in failure.

The Peking Man's story is, of course, much more complicated and much more interesting. When Dr. Johan Gunnar Andersson, director of Sweden's Geological Survey, came to China in 1914, he had been hired as a mining adviser to the Chinese government. Andersson was a self-described "mining specialist, a fossil collector and an archaeologist" who had led a Swedish survey in Antarctica from 1901 to 1903.[2] His arrival and interest in fossils, however, helped initiate a series of surveys and new modern research methodologies in northern China together with his Chinese and Swedish colleagues. The growing interest in Chinese history, prehistory, and paleohistory put China on a clear trajectory to becoming a major scientific force in archaeology and geology by the mid-twentieth century. "To many anthropologists in the 1920s, Asia seemed the most likely place for 'the cradle of mankind,'" offers historian Dr. Peter Kjaergaard. "Fame, prestige and money were intimately connected in the hunt for humankind's earliest ancestors and, thus, a lot was at stake for those involved. Several countries were competing for access to China as 'the paleontological Garden of Eden.'"[3]

Andersson's interest in China's fossils had been piqued by German paleontologist Max Schlosser's "dragon bone" findings from Schlosser's own travels in China over a decade earlier; by the time Andersson arrived in China, Schlosser's fossils had been taxonomically identified to ninety species of mammals. Many of the early fossil collectors came to find their specimens through Chinese locals who hunted the "dragon bones"—as fossils were called—as components for traditional medicines. Archaeologists went to dragon bone hunters for leads, suggestions, and the fossils the apothecaries had collected. In Schlosser's fossil collection was a humanlike third upper molar that piqued Andersson's interest in the area as a viable site for investigating human origins in Asia. For Andersson, this lone rogue tooth meant that there was clear evidence of early man in China; he just had to find it. From 1914 to 1918, Andersson paid a number of local technicians (or assistants, as he called them) for fossil hunting in the Shanxi, Henan, and Gansu provinces with the hope that some of these locales would successfully yield "dragon bones" or some other interesting objects of antiquity. Any materials that Andersson's lackeys recovered were promptly sent to Professor Carl Wiman of the Institute of Vertebrate Paleontology at Uppsala for study. In late autumn of 1920, Andersson's assistant Liu Chang-shan returned to Beijing with several hundred stone axes, knives, and other stone artifacts, all of which were from a single spot in the village of Yangshao in Henan.

Particularly significant about Andersson's work was his reliance on geological methods for excavating and his commitment to scientific methodology. "With geology, and the principles of stratigraphy as a means of exploring the dimension of time, there simply could be no scientific archaeology or any of its excavations that characteristically focus on the delineation of the context of objects," notes historian Dr. Magnus Fikesjö. "Andersson arrived . . . [at] his famous position at the beginnings of Chinese archaeology by way of his geology, precisely by observing stratigraphic patterns and scanning the landscape for traces of paleontological and human remains that might constitute new discoveries."[4] Artifacts were mapped to specific strata, and sites could be interpreted as a sequence of events with each of the excavated objects offering clues about what those events could have been. The reliance on geology's scientific framework firmly established the initial excavations—and the later excavations at Zhoukoudian—as credible modern science in China.

By 1918, Andersson's interest in fossils resonated with his colleagues, and J. McGregor Gibb—who was teaching chemistry in Beijing—showed Andersson some red-clay-covered fossil fragments from a place called Jigushan ("Dragon Bone Hill") near Zhou-koudian. (Zhoukoudian—also known as Chou Kou Tien or Choukoutien—was about forty kilometers from Beijing.) Andersson set out by mule on March 22, 1918, to explore the Zhoukoudian area, a day's travel from his home in Beijing. There, Andersson found a series of extensive limestone caves, with thick, fault-crossed sedimentary bands. Legend—coupled with oral history—has it that the Zhoukoudian area was first recognized as a fossiliferous locale as early as the Song dynasty (960–1279 CE), when archaeological evidence of limekilns appear in the area. Over millennia, the groundwater erosion of the limestone created caves and fissures, classic geomorphic catchment areas for the fabled dragon bones.

Andersson's initial exploration of the area reinforced his notion that it would be ideal for more systematic work, and in 1921 he assigned a young Austrian paleontologist, Otto Zdansky, to survey parts of the area. Zdansky, a recent graduate from the University of Vienna, had joined the team to collect fossils for Uppsala University. "I have a feeling that there lie here the remains of one of our ancestors and it is only a question of your finding him," Andersson gushed to Zdansky upon the latter's arrival at Zhoukoudian. "Take your time and stick to it till the cave is emptied, if need be."[5] Since Zdansky did not receive a salary for his work (although his expenses were covered), he had negotiated the rights to describe any fossil discoveries he made in the course of his work at Zhoukoudian. While Zdansky somewhat reluctantly began excavations at Zhoukoudian, Andersson turned his own attention to generating interest in the sites from other scientific institutions, organizing grants and donations and raising awareness about the site's significance. In one effort, Andersson brought Walter Granger, the chief paleontologist for an expedition underwritten by the American Museum of Natural History, to search for "early man." The plan was to garner Granger's notice about China's value to prehistory and the contributions China could offer the still developing scientific field, thus putting China's fossils squarely at the forefront of paleoanthropology's burgeoning interest in Asia.

During the 1921 field season, Zdansky unearthed that single tooth—a tooth with a worn-down crown and three roots. "Although Zdansky did not acknowledge the stone tools at Zhoukoudian as such," Kjaergaard argues, "[h]e soon realized that there was indeed ancient human remains buried at Choukoutien. However, he kept it to himself and put away the tooth he found. According to his own explanation, he did not want to let the sensation of a potential human ancestor cloud more important work. But, of course, he was perfectly aware of what this could mean for his career and what a compensation it would be for working without a proper salary."[6]

Zdansky did, however, deign to produce the tooth for visitors during the visit of the crown prince of Sweden to the site in 1926. But it wasn't until 1927 that the molar, in addition to another tooth fragment recovered from the excavation materials in the crates, was published by those working at the site. The tooth was identified as a molar from the right side of the mouth, from a species that Zdansky tentatively assigned to the genus *Homo*. (He put a question mark next to the species name.) Although Zdansky published his Zhoukoudian experiences in 1923—including a fossil catalog list of all the species recovered and identified—the questionable *Homo* tooth was conspicuously absent. After a subsequent field season in 1923, Zdansky returned to Uppsala and simply analyzed the tooth with the specimens recovered from his excavations. Although he bowed out of subsequent research at Zhoukoudian, the publication of a hominin tooth marked the beginning of a dedicated search for human ancestors at the site.

———

The presence of an "early man" or that elusive hominin ancestor—even if that presence was marked by just two teeth—was enough to motivate international agencies, like the Rockefeller Foundation, to fund excavations at Zhoukoudian. By 1927, the Rockefeller Foundation funds had arrived and systematic excavations began in earnest, under the leadership of Chinese scientists Dr. Ding Wenjiang (as the project's honorary director) and Dr. Weng Wenhao (later director of China's Geological Survey), as well as Canadian paleoanthropologist Dr. Davidson Black. Four scientific specialists—Dr. Anders Birger Bohlin (accompanied by his wife), Drs. Li Jie, Liu Delin, and Xie Renfu—were in charge of excavations and laboratory work. Other workers were hired, including a field manager and cook. Members of the field team stayed at the Liu Zhen Inn, a camel caravan inn that had just nine tiny damp adobe rooms. Located a mere two hundred meters from the site, it was rented by Li Jie for fourteen yuan a month and functioned as an ideal field headquarters between 1927 and 1931. Initial fieldwork began on March 27, 1927. Researchers conducted a systematic survey of the entire Zhoukoudian complex, extending to the county seat of Fang-shan, where earlier maps had been limited to only the Peking Man site proper. Full-scale excavations then began on April 16, 1927.

In addition to excavation funds, the Rockefeller Foundation sponsored the building and management of the Cenozoic Research Laboratory. Founded in 1928 by Davidson Black, Ding Wenjiang, and Weng Wenhao, the laboratory was a part of the Peking Union Medical College with the help of an $80,000 grant that Black had received from Rockefeller. The laboratory was specifically tasked to oversee the Peking Man material as the sheer amount of materials excavated and blasted out of the Zhoukoudian site boggles the mind: fossil specimens from the 1927 field season filled a staggering five hundred crates. Most of this fossil material was later shipped to the Museum of Far Eastern Antiquities in Sweden. (The transportation of the fossils from Beijing to Sweden wasn't without its own dangers; in November 1919, the Swedish ship *Peking* sank in a storm with eighty-two crates of plant and animal fossils on their way to Sweden for analysis. The loss of these fossils was a huge blow to the early days of Andersson's research.)[7]

By October 16, 1927, three days before the field season was supposed to end and as the team was beginning to wrap up their excavations, an in situ hominin tooth was discovered close to where Zdansky had found that tooth years earlier. In a letter dated October 29, 1927, Davidson Black wrote to Andersson, who was in Stockholm at the time:

> We have got a beautiful *human* tooth at last!
>
> It is truly glorious news, is it not!
>
> Bohlin is a splendid and enthusiastic worker who refused to permit local discomforts or military exercises to interfere with his investigation.... I couldn't get away myself for I was having daily committee work that demanded my presence here. Hsieh (Zie Renfu) couldn't reach Chou Kou Tien on account of local fighting. That night which was October 19th when I got back to my office at 6:30 from my meeting there I found Bohlin in his field clothes and covered with dust but his face just shining with happiness. He had finished the season's work in spite of the war and on October 16th he had found the tooth; being right on the spot when it was picked out of the matrix! My word, I was excited and elated! Bohlin came here before he had even let his wife know he was in Peking—he certainly is a man after my own heart and I hope you will tell Dr. Wiman how much I appreciate his help in securing Bohlin for the work in China.
>
> We have now in Peking some 50 boxes of material which we got in last July when the last military crisis was on but there are 300 more large boxes yet to come from Chou Kou Tien. Mr. Li of the Survey is busy trying to get rail cars to bring back this material. It will fill more than two cars![8]

His enthusiasm was well placed. Before 1929, the excavations at Zhoukoudian had resulted in only a few more isolated hominin teeth—not a lot more than what had been recovered between 1921 and 1927. The 1929 field season saw the beginning of excavations in the middle part of the Zhoukoudian deposits—these deposits were west of the northern fissure that crossed the site. The 1929 field season proved to be a real turning point in the Zhoukoudian excavations, in large part due to what was uncovered in December of that year. "Dragon Bone Hill"—also referred to as "Locality 53" in Andersson's early notes—was renamed "Cave 1" and appeared as such in all subsequent documentation. The project paid a yearly rent of 90 yuan to a coal company (the hill was a quarry site the company owned), which was raised to 180 yuan after 1927. To prevent what excavators considered "extortion," the Cenozoic Research Laboratory paid the "exorbitant" price of 4,900 yuan for permanent use of the site.

Where Bohlin and Li Jie had shared administrative and scientific affairs, paleontologist-anthropologist Professor Pei Wenzhong had to deal with the overwhelming logistics of running such a large site by himself; in interviews decades later, Pei recalled that he was seized by melancholy after Black left in April 1929 and Pei took over the site's care.

By November 1929, the site proved to be extremely rich in fauna—145 antelope jaws were excavated in one day, for instance. The cache of antelope joined the faunal record with complete pig and buffalo skulls, as well as antlers—yet few hominin teeth. In the late afternoon of December 2, 1929, however, the story of human origins in China found itself with a new fossil character. Workers discovered a skullcap in the fifth stratigraphic layer at Zhoukoudian; the presence of the cranium was clear, irrefutable evidence that the story of human evolution had early ties to China. The discovery of these fossils meant that China's rather recent use of geologic methodology and science was able to juxtapose itself with a strong commitment to "Chinese" historical antiquity, giving the entire excavation a nationalist agenda. With this one discovery, Chinese history pushed its antiquity and "legitimacy" back to the Pleistocene and became a serious force within the developing science of human origins.

Archaeologists and workers excavating at the Peking Man discovery site, China, 1920s. *(Science Source)*

The sheer excitement of the discovery was unmistakable. In a series of interviews in 1980, Pei Wenzhong recalled the details from December 2, 1929:

> In the afternoon after four o'clock, it was near sunset and the winter wind brought freezing temperatures to the site. Everybody felt the cold, but all were working hard at finding more fossils. . . . The large number of fossils attracted everyone of us and we all went down to take a look, so I know what it was like down there in the crack.

We generally used gas light, for it was brighter. But the pit was so small that anyone working there had to hold a candle in one hand and work with the other.[9]

———

Prehistorian and archaeologist Dr. Jia Lanpo offered his recollections of the discovery of the first Peking Man skull:

Maybe because of the cold weather, or the hour of the day, the stillness of the air was punctuated only by occasional rhythmic hammer sounds that indicated the presence of men down in the pit. "What's that?" Pei suddenly cried out. "A human skull!" In the tranquility, everybody heard him.

Pei had gone down after the sighting of the fossils, and now, where he was told there was a round-shaped object there, he had stayed there and worked with the technicians. As more of the object became exposed, he had cried out. Everybody around him was excited and gratified at the long-awaited find.

Some suggested that they take it out at once, while others objected for fear that, working rashly in the late hours, they might damage the object. "It has been there for so many thousands of years, what harm would it do lying there for one more night?" they argued. But a long night of suspense was too much to bear.[10]

Pei's terse telegraph to Black fittingly captures the emotion of the moment: "Found skullcap—perfect—look[s] like man's."[11] The news was scarcely believed at first; skeptics either doubted Pei's ability to correctly identify the fossil specimen or, after two years of excavations with only the occasional tooth to show for their efforts, refused to believe that the excavations could have been so lucky. In a letter to Andersson dated December 5, 1929, Davidson Black wrote: "I had a telegram from Pei from Chou Kou Tien yesterday saying he would be in Peking tomorrow bringing with him what he thinks is a complete *Sinanthropus* skull! I hope it turns out to be true."[12]

Simply finding the fossil wasn't enough, though. The specimen had to be carefully excavated and transported to the Cenozoic Research Laboratory. Excavating and storing the fossil was a bit tricky; when that *Sinanthropus* fossil was first unearthed, the specimen was rather wet and soft, due to the cave's sediments, and could be damaged easily. The specimen thus had to be dried out before it could be transported to Beijing. Pei and fellow archaeologists Qiao and Wang Cunyi stayed day and night next to a fire to dry out the skull. Pei carefully wrapped it in layers of gauze; the gauze was covered with plaster and dried again; then it was all wrapped in two thick cotton quilts and two blankets and the entire

Excavations at Zhoukoudian showing how fossils were jacketed in situ for safe removal. From Paramount News film, early 1930s. *(Film courtesy of the American Museum of Natural History Library and Dr. Milford Wolpoff)*

specimen was then trussed up with rope. What had been so carefully excavated from one set of soil and cave sediments was now jacketed in a new stratigraphy of cultural layers and materials. Pei delivered the first complete skull of the Peking Man assemblage to Davidson Black at the Cenozoic Research Laboratory on December 6, 1929.

China's Geological Survey held a special meeting on December 28, 1929, to announce the discovery; the next day the foreign press reported the news of the phenomenal fossil find, which quickly spread across global scientific communities. Scientists like British anatomist Grafton Elliot Smith—while still immersed in sorting out the anatomy of Piltdown Man—visited Beijing in September 1930 to examine the Peking Man fossils. Over the next few years, the Cenozoic Research Laboratory continued its excavations at the Zhoukoudian sites, and additional fragments of skulls, jaws, and teeth were recovered; all were assigned to *Sinanthropus*.[13]

On March 16, 1934, Davidson Black passed away—he was found dead that morning with the Zhoukoudian specimens lying in front of him as he attempted to catch up with work. Dr. Franz Weidenreich, a German anatomist, took over Black's position and work in 1935. Weidenreich's attention to detail and scientific brilliance helped push the scientific import of the Zhoukoudian fossils to the forefront of the scientific community. Unfortunately, Weidenreich was not as sociable and personable as his predecessor. He left all administrative organization and affairs to his Chinese counterpart in the laboratory, Yang Zhongjian, who had earlier directed the excavations at Zhoukoudian, from 1928 to 1933. As a result of Weidenreich's reticence about administrative affairs, the Rockefeller Foundation stopped supporting the Cenozoic Research Laboratory directly, although the Foundation continued to finance the Zhoukoudian excavations, allocating money for continued work there through March 31, 1937. In meeting minutes, the Rockefeller Foundation acknowledged the incredible scientific significance that the site offered both China and the international community:

> The paleontological finds in the caves of Choukoutien near Peking constitute one of the most dramatically interesting and significant advances ever made in our knowledge of ancient man. The scientific importance of this work can not be questioned, and the collapse of the program would be a major scientific loss. The program has, moreover, been closely associated from the outset with the Peiping Union Medical College. It represents a fine-spirited cooperation between Chinese and western scholars and in terms of scientific competence and achievement it is outstanding in China's experience. It was natural to fear that Dr. Black's death would mean the virtual end of this project. However, Dr. Franz Weidenreich, formerly of the University of Frankfurt and of the University of Chicago, has, since his appointment in March, 1935, demonstrated the necessary qualities of scholarship, administrative ability, tact, etc., to carry forward with distinguished success the work which was so brilliantly begun by Dr. Black.[14]

Despite this endorsement, the beginning of the Second Sino-Japanese War in 1937 and the difficulties that the war raised meant that excavations at Zhoukoudian stopped and fossils were carefully locked away in the laboratory. Weidenreich was concerned that if Japan and the United States went to war, the Japanese would take over the lab; in the summer of 1941, Weidenreich insisted that additional replicas of the bones be created. In late 1941, Weidenreich left Beijing, opting to take a position at the American Museum of Natural History.

So what made Peking Man "Peking Man"? Taxonomically, Peking Man was part of a species that Johan Andersson and colleagues named *Sinanthropus pekinensis*—not a single individual but a series of individuals now known as *Homo erectus*. Davidson Black's initial morphological studies described a species similar to modern humans, having a large brain but overall similar skull and bone sizes. *Sinanthropus*, however, was different in that it had heavy brows and large, chinless jaws. Geologically, the site dates to between 750,000 and 530,000 years ago. Today, thanks to extensive analyses of the site's artifacts, we know that the species had sophisticated stone tools and offered the first systematic use of controlled fire outside of our own species, *Homo sapiens*. From a historical standpoint, however, the moniker "Peking Man" refers to the assemblage of fossils found at Zhoukoudian. When we talk about "Peking Man," we are thus implicitly referring to both a taxonomic moment in time and the identity of a historical object.

"There is a celebrity around the fossils, especially in the 1920s and 1930s, when they become quite individualized and personalized," historian Dr. Christopher Manias explains. "You do get the sense that the media or popular accounts are talking about 'Peking Man' as a definite individual and trying to work out what 'he' was like: who he was, when he lived, what moral standard he was at, what he ate, how much like 'us' he was, and so on."[15]

Where other fossil discoveries had clear nationalistic ties—the Piltdown Man, for example, was touted as "the earliest Englishman"—no other discovery was quite as inexorably linked to the development of science writ large in the way that Peking Man was. Many standard histories regard the development of modern geology in China as influenced by foreign imperialism, with only a few Chinese students studying abroad in the West and then returning to China in the early to mid-twentieth century, bringing back with them Western techniques and theories. This was different than science done in British colonies, and to that end China had a different kind of paleoanthropology than the science that surrounded the Taung Child in South Africa. One of the reasons that the European paleointelligentsia was so sniffy about Taung's discovery was that the fossil and the fossil's ancestral interpretation had come from a colony—South Africa—and they felt it ought to have been validated by the European (specifically British) establishment.

The introduction of the scientific methodology and framework offered a means of legitimizing China's presence and participation in global scientific norms of geology. "To China's geological pioneers, the connection between nation and science was even more basic. Whether they were collecting rocks and fossils or elucidating earth processes, they were in a sense studying China directly and fitting it into a global narrative," argues historian Dr. Grace Yen Shen.[16] Participating in these new frameworks of geology and archaeology became a way for China to take part in the global modernity of geological sciences, and China was now a serious player in paleoanthropology. It's hard, in fact, to imagine a more global perspective than a search for "early man"—which was, after all, how the Zhoukoudian project unfolded. It was backed by a variety of international participants, and the site and its treasures meant that *Sinathropus's* early identity was an international one. Workers at the site were Canadian, Swedish, Austrian, German, and French as well as Chinese. Moreover, the excavation of Zhoukoudian tapped into a variety of scientific, intellectual networks devoted to studying archaeology, human evolution, and the *longue durée* of human history, with international ties due to the long-running Swedish connection, as well as a French involvement.[17] Even with this international focus, the fossils themselves became a strong symbol of China and its history.

———

While the specific moment of the discovery of the Zhoukoudian fossils is a bit ambiguous, the date of the fossils' disappearance is specific—but circumstances, even decades later, are far from clear. And as with most aspects of Peking Man, there is a long version and a short version to the story.

In the short version, researchers at the Peking Union Medical College, especially Franz Weidenreich, were concerned about the safety of the fossils due to building tensions between China and Japan between 1939 and 1941. When the United States declared war on Japan on December 8, 1941, after the bombing of Pearl Harbor, the Japanese military took over the Peking Union Medical College. Concerned that the fossils would be looted from China—or completely destroyed—the College carefully crated the Peking Man fossils with the intention of smuggling them out of China to the United States or Europe. The fossils were packed into two crates and driven to the U.S. Marines' base at Camp Holcomb, where they were scheduled to be shipped out on the USS *President Harrison*. However, the fossils happened to arrive at the base just a few days before the U.S. military base surrendered to the Japanese. Somewhere between the departure of the fossils from Beijing and loading on board the *Harrison*, the fossils were lost in the confusion and pandemonium.

The long version of the Peking Man disappearance reads like something straight out of a Dashiell Hammett novel—there's mystery and intrigue, some fact but more fiction. It's as if the hardboiled detective Sam Spade has been tasked to track down priceless scientific curios.

In preparation for transport, anatomy technicians Ji Yan-qing and Hu Chengzhi of the Cenozoic Research Laboratory wrapped each fossil in white tissue paper, cushioned them with cotton and gauze, then overwrapped the fossils with white sheet paper. The fossils were placed in small wooden boxes with several layers of corrugated cardboard on all sides. The smaller wooden boxes were then placed into two big unpainted wooden crates, one of which was approximately the size of a large office desk and the other slightly smaller. The crates were delivered to Controller T. Bowen's office at the Peking Union Medical College. Once the Japanese military attacked Pearl Harbor and the Japanese took over the College, the fossils in their crates were moved around different storerooms, then quickly delivered to the U.S. embassy at Dong Jiao Min Xiang in Beijing. All of this occurred three weeks before the attack on Pearl Harbor.

The contents of the two crates reflected the vast amount of archaeological materials that had been excavated at the Zhoukoudian site. Case 1, for example, had seven boxes nestled into the desk-sized crate: Box 1 contained teeth (in seventy-nine separate smaller boxes), nine thighbone fragments, two fragmented humeri, three upper jaws, a collarbone, a carpal bone, a nasal bone, a palate, a cervical vertebra, fifteen skull fragments, a separate box of skull fragments, two boxes of toe bones, and thirteen boxes of mandibles. Case 1 also contained six additional boxes of skulls and a small container of orangutan teeth. With the exception of the orangutan teeth, all of the fossils in Case 1 were assigned to Peking Man, indicative of how Peking Man, with thirteen jawbones and nine thighbones, reflected an assemblage of multiple individuals that represented all genders and ages. The second crate contained a similar swath of Peking Man fossil remains, plus several macaque (monkey) skulls. The lab took careful notes about these crates as well, noting who packed which one and what kind of packing materials were used. Although these crates were also lost and never recovered, the notes about their contents have survived.

Backing up, however, to the buildup of political and military tensions in November and early December 1941, Dr. Weng Wenhao, the director of China's Geological Survey, appealed to Dr. Henry Houghton, president of the College, to have the Peking Man collection taken to safety. Houghton asked Colonel William W. Ashurst—a commander of the marine detachment at the U.S. embassy in Peking—to send the Peking Man collection to safety, under protection of the marines, leaving within a couple of days. At five a.m. on December 5, the marines' special train—with the Peking Man fossils—pulled out of Peking, headed down the Japanese-owned Manchuria railroad toward the small Chinese coastal town of Chinwangtao. From Chinwangtao, the Peking Man materials were to be loaded onto the American liner USS *President Harrison*, which was to head to Shanghai, then north from there.

However, the Japanese attack on Pearl Harbor halted all plans. To prevent the capture of the *Harrison*, her crew grounded her at the mouth of the Yangtze River, and the marine train with the fossils was captured by the Japanese at Chinwangtao. What happened to the two large crates of Peking Man remains has been the subject of much speculation, due in no small part to the varied and often contradictory testimonies of different witnesses. "What happened from that moment on is clouded in rumour and the confusion of war," author Ruth Moore writes in her account of the Peking Man disappearance. "Despite the efforts of three governments to find them, they have vanished from the world as completely as during the centuries when they lay hidden in the earth of Dragon Bone Hill. According to one account, the Japanese loaded all the cases taken from the train on a lighter that was to take them to a freighter lying off Tientsin. The lighter, it is said, capsized, and the remains of Peking Man drifted away or sank to the bottom of the sea. The other story is that the Japanese who looted the train knew nothing of the value of the scraps of bone and either threw them away or sold them to Chinese traders as 'dragon bones.' If so, they may long since have been ground into medicine."[18]

––––––––

Since the Peking Man story lacks a satisfying resolution, many believe that the fossils are still out there just waiting to be rediscovered, and such believers have mounted decades of searches.

This is where, in 1972, one Christopher Janus—a U.S. financier and philanthropist from Chicago—enters the Peking Man story. Janus was no stranger to public outrage, having himself owned and driven Hitler's limousine. Moreover, in 1950, he'd inherited a cotton plantation and "fifty Egyptian dancing girls," whom he used as a vaudeville act; the exasperated Egyptian embassy spent months explaining that slavery was outlawed in Egypt and desperately tried to distance themselves from Janus, whom they had come to see as a political leper.

Like a character straight out of a film noir (and with a name to match), Janus was determined to write his own chapter of the Peking Man story. The disappearance of the fossils had attracted his interest during his visit to China in 1972 among the first group of Americans allowed to visit the country as it reopened to the West. His dynamic personality was matched by his penchant for history and interest in culture. Although Janus had no training as an anthropologist—he admitted that he had never even heard of the Peking Man fossils before his trip and his visit to the Peking Man Museum—Janus felt that he had been selected and charged by Dr. Wu of the Peking Man Museum at Zhoukoudian to find the fossils and return them to China. To read how Janus tells his story, the return of the Peking Man fossils turned into a personal mission. Upon his return to the United States, Janus quickly set to finding the missing specimens, offering a $5,000 reward for information about their whereabouts.

His book, *The Search for Peking Man*, teems with mystery and conspiracy—clandestine meetings, cloak-and-dagger innuendo, and international intrigues. The first couple of chapters of the book describe the loss of the fossils and relate in vivid detail how a Dr. Herman Davis supposedly used the boxes as poker tables. According to Janus's "research," Davis even used the crates of fossils to steady his machine gun during the Japanese invasion of the base. Janus has people coming out of the woodwork to offer their take on the possible fate of

Peking Man: some claim to know where it is; others claim to actually have the remains. For example, Mr. Andrew Sze, a Chinese expat living in the United States, claimed that the fossils were in Taiwan and his best friend knew the exact location. The climax of Janus's search involved a particularly furtive meeting with a woman who claimed to have the fossils in her late husband's U.S. Marine footlocker—he had brought the fossils back from his deployment in World War II, she said. Janus was to meet the woman at the top of the Empire State Building at noon one spring day; she told him that he would know her because she would be wearing sunglasses. On the rooftop, she gave him a blurry photo of what looked like fossils, then utterly vanished. (Harry Shapiro, of the American Museum of Natural History, was dubious at best about the materials when Janus asked him to look at the photo to authenticate the fossils. The photograph in question is spectacularly blurry and conveniently out of focus.) Janus also claimed his search for the fossils continued—rather improbably—with help from the FBI and CIA, claiming to want to help him locate the fossils "in the national interest."[19]

Janus's hunt for Peking Man came to a screeching halt on February 25, 1981, when he was indicted by a federal grand jury on thirty-seven fraud counts. Prosecutors charged that the international search for the bones amounted to a $640,000 fraud in which Janus had funneled the majority of the funds—$520,000 in bank loans and $120,000 from investors to finance the search and produce a film—toward his personal use. In an interview with the *Chicago Tribune*, Janus insisted that all the money he borrowed was for the search and the planned film. After his indictment, Janus hinted that U.S. relations with China would be ruined if the federal government took action against him. "The whole thing is more than the search for the Peking Man," Mr. Janus told the press. "It involves certain relationships with China that can't be discussed, a project we have going with the Federal Government."[20]

The grand jury concluded that Mr. Janus had made no serious effort to search for the Peking Man or to make the film. But it could not find out what he did with most of the money he'd borrowed. "He'd say, 'I see Harrison Ford as me,'" recalls William Brashler, coauthor of *The Search for Peking Man*, in an interview with the *Wall Street Journal*. "He immediately hit me up to invest in the movie. It was hard not to like him, but he had one arm around your shoulder and the other in your wallet."[21] Ultimately, Janus pled guilty to two counts of fraud.

Where characters like Janus found blatant means of inserting themselves into the Peking Man story, others, like Claire Taschdjian, a technician at the Peking Union Medical College and one of the last people to have seen the fossils, participated in the Peking legacy in a more subtle way. Taschdjian wrote *The Peking Man Is Missing*—a fictionalized account of the fossils' disappearance. (The book can most charitably be described as sensationalistic—full of torpid prose, kept together by a hilariously simplistic plot.) But Taschdjian was a secretary at the laboratory in Beijing when the fossils were lost, and by a quirk of historical happenstance, her comments on the fossils— and anything she writes—have a tell-all sensationalism to them since she was one of the last people to have seen the actual fossils. In January 1975, the original *Hawaii Five-O* ran an episode, "Bones of Contention," in which Steve McGarrett's team tracks down the "world's oldest missing persons case"; they find the remains of Peking Man in a military storage unit in Hawaii. It's the thrill of the hunt, the treasure, and the mystery that drives the fiction. And it's that very sensationalism that cuts to the core of how we are geared to think about the Peking Man story. The fossils' fame hinges now on the mystery and intrigue that surround him; it's only logical, then, that the stories we create and repeat about the fossil end up just as romanticized as the fossils themselves.

Even as recently as July 2006, Beijing's Fangshan district government announced that it was renewing its search for the fossils. A committee of four from the museum located on the Zhoukoudian site began to gather leads for the fossils' whereabouts throughout China. A search hotline was even published in the local newspapers; by the fall of that year, the committee announced that sixty-three total leads had come in. One committee member, quoted in multiple newspapers, said that four leads looked "especially promising." Lead number one: A "121-year-old man" who had served as a high official in Sun Yat-sen's republican government said he knew exactly where the fossils were. Lead number two: An "old professor" from northwest Gansu Province, during a visit to Japan, had found revealing testimony from an American soldier in the Tokyo military tribunals' archives. Lead number three: A Mr. Liu, from Beijing, said he knew an "old revolutionary" who had a skull in his possession. Lead number four: Another Beijing man said that his father, a former doctor at Peking Union Hospital, had brought one of the skulls home from work one day and buried the fossils in his neighbor's yard.[22]

None of these leads panned out.

———

If the fossils are missing, how can Peking Man have any kind of scientific legacy? In the first part of the twentieth century, casts of fossil specimens were the key to paleosciences. Since fossils were too valuable and rare to ship to international researchers, casts of fossils were sent through networks of natural history institutions. (Recall that Raymond Dart had specifically insured the Taung Child for marine travel when he sailed to London from South Africa.) In the early days of human origins research, paleoanthropologists would offer to trade casts of "their" fossil to other researchers in different areas of the world who had different specimens—the casts thus became a kind of social currency. Scientific colleagues—both collaborators and detractors—wanted to see copies of the fossil in order to examine its anatomy for themselves. People outside of academic circles had heard of the famous fossils and expected to see them in public museums. In order to circulate them for study and display, accurate copies of the fossils had to be made.[23]

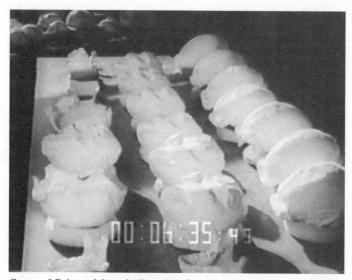

Casts of Peking Man skull at the Cenozoic Research Laboratory, curing and drying on laboratory bench. From Paramount News film, early 1930s. *(Film courtesy of the American Museum of Natural History Library and Dr. Milford Wolpoff)*

"All [*Sinanthropus pekinensis*] casts are made and coloured with extreme care and attention to the finest detail. They can be studied with complete confidence," advertised the catalog for R. F. Damon & Co., purveyor of fossils and creator of fossil casts.[24] With the company's new catalog page for the fossils excavated during Zhoukoudian's field seasons of the early 1930s, access to Peking Man was suddenly available to international researchers. Every scientist from Sir Arthur Keith in London to Raymond Dart in South Africa could examine the remarkable Zhoukoudian finds.

To that end, on August 2, 1930, Pierre Teilhard de Chardin wrote to Marcellin Boule about the exciting discoveries at Zhoukoudian and Teilhard de Chardin's own studies in working out the comparisons between different fossil taxa. A powerful presence in the early-twentieth-century paleo world (having worked with both the La Chapelle Neanderthal as well as the Piltdown fossil), Teilhard de Chardin shifted the focus of his work to China when excavations began at Zhoukoudian. "On returning to Peking, I had the pleasant surprise of finding at Black's laboratory a second skull of *Sinanthropus*, identical to the first by form and also (fortunately) by its state of conservation. In this second sample, one discerns the beginning of the nasal bones, and some further details," Teilhard wrote. "Black has made some casts (very good ones) of all the isolated pieces. Two weeks from now, he should be able to give an estimate of the cranial capacity, taken from one absolutely perfect piece as preparation."[25]

Although they made the exchange of scientific information easier, the casts represented a huge commitment of time, resources, and investment. "Casts preserve the external form of the fossil, and they thus represent a permanent duplicate record of the shape of fossil bones. They are routinely used in place of original fossils for research, since they enable scientists to study and compare the remains of animals that have been discovered thousands of miles apart, and may be stored on different continents," museum curators Drs. Janet Monge and Alan Mann explain. "Virtually every paleontological museum and academic department spends considerable time in the procurement of quality casts for both research and instructional purposes."[26]

By 1932, when R. F. Damon & Co. was expanding its collection of *Sinanthropus* casts, Robert Ferris Damon inherited the company from his father, Robert Damon. Damon Senior had established himself in the fossil business in 1850, undertaking the artistic and technological aspects of creating good fossil casts for paleontologists and prehistorians. All of these casts were made of heavy plaster and were used by museums in collections as well as in displays. In the early days of the casting company, between 1850 and 1900, most of the casts and models were marine shells and fish. With the influx of hominin fossils, the company expanded its paleontological collections to include anthropological casts and models. As interest burgeoned in obtaining casts of human ancestors and anthropological specimens, the company focused on skulls, jaws, and teeth of humans and their ancestors. With the discovery of fossil hominins in Southeast Asia in 1891 (Java Man), Europe in 1912 (Piltdown), and Africa in 1924 (Taung Child), many researchers and museums wanted access to copies of fossils to be able to examine the specimens for themselves.

In the mid-1930s, R. F. Damon & Co. was authorized by Davidson Black and Weng Wenhao to expand the list of casts available of *Sinanthropus pekinensis* as more and more specimens of Peking Man came from the excavations. These new casts included eight mandibular fragments from a variety of differently aged individuals, juvenile through adult, to a skull from Locus E and based on the materials from Davidson Black's 1931 publication in *Palaeontologia Sinica*. Prices for such materials were usually several pounds.

Without any originals whatsoever, researchers are left with only the casts as the material evidence and tangible remains of the early Zhoukoudian excavations. Where other casts merely carry the information of the original fossils, the Peking Man casts have come to take the place of the originals. "Fortunately, during the time when they were studied in China, quality plaster casts of almost all of the Zhoukoudian bones were made and distributed to major museums around the world," Monge and Mann note. "These casts preserve a remarkable amount of detail, and in many cases, measurements taken from them show no significant difference from measurements recorded on the original fossils. This represents a remarkable achievement considering the level of molding and casting technology in the 1930s, and the (by today's standards) primitive molding media.

NEW CASTS OF
SINANTHROPUS PEKINENSIS.

MESSRS. R. F. DAMON & Co. have been authorised by Prof. DAVIDSON BLACK, F.R.S., of Peiping Union Medical College and Dr. W. H. WONG, Director of the Geological Survey of China, to make the following additions to their list of casts of Sinanthropus pekinensis.

References : SKULL—Palaeontologia Sinica, Series D., Vol. VII., Fascicle 2, 1931.
JAW SPECIMENS—Pal. Sin., Ser. D., Vol. VII., Fasc. 3, 1933.

			£	s.	d.
544	Locus B jaw.	Juvenile. Right ramus and symphysis in stage of preparation showing errupted and unerrupted permanent dentition ...	1	12	6
545	Locus B jaw.	Symphysial region	1	1	0
546	Locus B jaw.	Entirely freed from matrix and restored	1	15	0
547	Locus A jaw.	Adult right ramus with 1st, 2nd and 3rd molars and sockets of premolars and canine	1	10	0
548	Locus F jaw.	Juvenile. Posterior portion of right ramus with 1st permanent molar errupted and 2nd permanent unerrupted molar exposed	1	10	0
549	Locus C jaw.	Fragment of adolescent, unerrupted 3rd molar exposed ...	1	1	0
550	Locus G.1 jaw.	Left ramus with complete permanent dentition ...	1	15	0
551	Locus G.2 jaw.	Ascending right ramus with 2nd and 3rd permanent molars in situ	1	12	6
552	Terminal phalanx	(Bull. Geol. Soc. China., Vol. XI., No. 4)		10	0
553	Locus E skull.	Complete as shown in Pal. Sin., Ser. D, Vol. VII., Fasc. 2, 1931, Pls. XI., XII., XIII., XIV.	4	15	0
554	Locus E skull.	Endocranial cast	2	2	0

For Sinanthropus casts already offered, see separate list
(Nos. 510, 511, 529, 530, 531, 532, 533).

All casts are made and coloured with extreme care and attention to the finest detail. They can be studied with complete confidence.

R. F. DAMON & Co.,
45, HAZLEWELL ROAD, LONDON, S.W.15.

PLEASE NOTE.—*Orders will be executed in rotation as received.*

Brochure advertising Peking Man casts from the prominent R. F. Damon & Co. *(Raymond Dart Collection. Courtesy of the University of the Witwatersrand Archive)*

Although no cast is an ideal substitute for the original fossil, in this case, the casts represent the *only* record of these specimens, and provide a reasonable alternative for the missing originals."[27]

In 1951–1952, when China was actively looking to have the original Peking Man fossils returned, the casts were confused with the original specimens. In a letter dated October 6, 1951, Dr. Walter Kühne, a paleontologist at Humboldt University in Berlin wrote to Yang Zhongjian, the director of the Institute of Vertebrate Paleontology and Paleoanthropology in Beijing. In his letter, Kühne claimed to have been told by a colleague, Dr. D. M. S. Walson, that Walson had seen skullcap 2 of Zhoukoudian at the American Museum of Natural History in New York and, further, that Walson had seen Weidenreich himself handling the specimen. This claim about the fossils immediately sparked an editorial in the *People's Daily* (dated January 1, 1952) that urged the American Museum of Natural History—and, indeed, the United States—to return the fossils to the People's Republic of China. However, in a letter dated April 29, 1952, Dr. Joseph Needham, president of the Britain-China Friendship Association, proved that Walson was mistaken about the identity of the fossil and included a letter from Dr. Kenneth Oakley (of Piltdown debunking fame) that demonstrated that what Walson had seen was, in fact, merely casts. Walson himself retracted his claim once his error had been pointed out.[28]

What does it mean, then, for us to be left with the replica of a fossil? Does it even matter? "Even without the originals, the duplicates of the Peking Man fossils made before their disappearance have provided substantial information for morphological studies of *Homo erectus*," historian of science Dr. Hsiao-pei Yen claims. "Therefore, it is questionable if the discovery of any of the missing Peking Man original fossils would dramatically change our current understanding of human evolution."[29]

On one level, this sentiment is certainly true. If the fossils are simply their dimensions and their physical forms, then certainly Yen is correct that the casts are just as good as the originals. Yet the original fossil clearly carries with it cachet and cultural value beyond its height and breadth; in this sense, such an argument amounts to saying that a copy of the Hope Diamond or the *Mona Lisa* would be the "same" as the original.

Painted casts of Peking Man skull at the Cenozoic Research Laboratory on a laboratory bench. From Paramount News film, early 1930s. *(Film courtesy of the American Museum of Natural History Library and Dr. Milford Wolpoff)*

"In 1539 the Knight Templars of Malta paid tribute to Charles V of Spain, by sending him a Golden Falcon encrusted from beak to claw with the rarest jewels—but pirates seized the galley carrying this priceless token and the fate of the Maltese Falcon remains a mystery to this day," reads the introductory text that appears after the opening credits in the 1941 film *The Maltese Falcon*. It is the story of a treasure hunt for a priceless object and the motivations that fuel that hunt. The "black bird" that Kasper Gutman and Sam Spade search for is that jewel-encrusted falcon, which, by the 1940s, was said to have been covered in a deep black patina to hide the bird's true value. In the film's dramatic reveal—where the bird is proven to be a fake—the audience is told that the bird is more myth than fact. In the end, actually finding the bird wasn't as important as cultivating the belief in what it stood for. The bird, Sam Spade dryly notes, is the "stuff that dreams are made out of."

Today, the only pieces of the Peking Man in Chinese collections are five teeth and some parts of a skull found in the renewed excavation of the 1950s and 1960s. The Uppsala Museum of Evolution has three teeth from the original excavations in the 1920s; these are considered the "collection's highlights." When the tooth was discovered in the boxes of Professor Carl Wiman's stuff, that tooth, newly reexcavated from the archives, became a significant part of the Peking Man's story. Like that tooth, Peking Man's story is one of abrupt beginnings and endings, encounters and losses; it's a story of details and dramatic events—kind of like *The Maltese Falcon*, but with fossils.

"As is well known, almost all material from this excavation period (except the original Uppsala teeth) was lost in 1941, and has never been recovered," Swedish research Dr. Per Ahlberg said in an interview. "After the war, Chinese scientists continued to excavate Zhoukoudian and found some new fossils in the deeper layers. But this new tooth is most probably the last fossil from the 'classical' Peking Man excavations that will ever be found." Ahlberg continued, "We can see many details that tell us about the life of the owner of the tooth, which is relatively small, indicating that it belonged to a woman. The tooth is also rather worn, so the person must have been rather old when she died. Also, parts of the tooth enamel have been broken off, probably indicating that the person had bitten down on something really hard, like a bone or a nut. We should probably now be talking about a 'Peking Woman' and not a 'Peking Man.'" Professor Liu Wu, from the Chinese Academy of Sciences, chimed in with his observations that the canine tooth was fractured but otherwise well preserved: "This is an extremely important find. It is the only canine tooth in existence. It can yield important information about how *Homo erectus* lived in China."[30]

Peking Man's legacy—his legend, his fame—hinges on his disappearance. It's as if the paleo world had found a historical parallel for the story of Amelia Earhart; Peking Man captivates its audiences

Peking Man teeth with
original museum label
(Lagreliska Collection).
(Science Source)

because the ending of its story is a mystery. As history, unresolved
stories can be unsettling and deeply unsatisfying. Even Piltdown—
with his conspiracies—is a fossil with a better-resolved narrative.
Piltdown is a hoax; the perpetrator might still be at historical large,
but the fossil's story has been rather neatly tied up with Kenneth
Oakley's chemical analysis, and Piltdown's fossils are carefully stored
in the fossil vault at the London Natural History Museum. Peking
Man, on the other hand, is missing—it's a paleo-noir cold case in the
history of science.

Perhaps the black bird offers a useful lens for making sense of
the life history of the Peking Man fossils. Every aspect of the Peking
Man story contains multiple levels. One is its science, of course, but
also finding part of the Peking Man collection—even if it is just a

tiny part found in archival boxes—provides a compelling narrative aspect, another level, to the Peking Man story. What was lost has now been found.

Today, we know Peking Man through the recently recovered canine tooth and a couple of other molars sent back to Uppsala from Zhoukoudian during those initial excavations—but we know Peking Man better through his plaster casts, photographs, and stories. The fossils are famous because we don't have them anymore. Peking Man is, indeed, a curious case of celebrity; a fossil made famous by its paleo-noir mystique.

To date, the fossils have not been recovered.

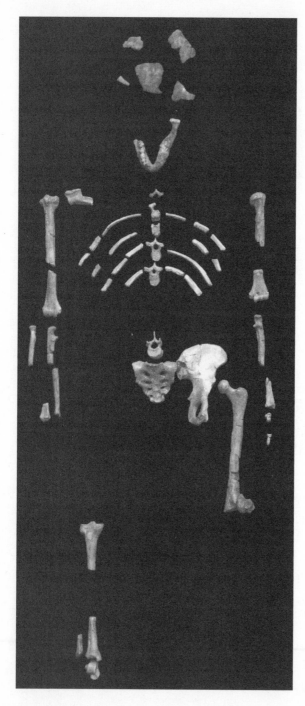

Portrait of Lucy,
AL 288-1.

CHAPTER FIVE

THE ASCENSION OF AN ICON:
LUCY IN THE SKY

T o Locality 162 with Gray in AM. Feel good," paleoanthropologist Dr. Donald Johanson wrote in his field journal
early the morning of November 30, 1974.[1] The good feeling Johanson recorded in his notes turned out, in fact, to be a lucky
premonition—that morning, he and graduate student Tom Gray discovered an amazing fossil on the slopes of Hadar, a fossil site in
northern Ethiopia where Johanson's team had been working that
field season. Johanson and Gray's remarkable discovery was a fossil
hominin skeleton, the most complete early hominin skeleton in
paleoanthropology's collections. The team promptly christened the
discovery Lucy. In subsequent years, researchers assigned Lucy to
a then new species of extinct hominin, *Australopithecus afarensis*—a
species that lived 3.25 million years ago. In the years since her discovery, she has become one of the twentieth century's most iconic fossil finds.

Some fossils acquire special significance or celebrity status because
they are a "first" of something, or the oldest, the original, or even the
place with the most of a unique collection of fossils. A few fossils

become archetypes for representing long-extinct species, like the Old Man of La Chapelle. Some become type specimens, significant in their standing as a prototype of a fossil's biological category, like the Taung Child. Some fossils can be cultural emblems that crystallize a particular tradition of scientific thought and practice, and directly impact the trajectory of future research. But the discovery of Lucy that November morning in 1974 introduced a new kind of celebrity fossil to paleoanthropology. It introduced a fossil that would function as an icon—a venerated scientific object, a key piece in the puzzle of human evolution, and a cultural touchstone. In the span of four decades, Lucy has moved from merely being an extinct taxonomic classification of a then species of hominin, *Australopithecus afarensis*, to being the specimen against which all fossils are measured.

———

To make a fossil discovery famous, it needs a compelling origin story, and Lucy's narrative is masterfully constructed. Just like the episode of Raymond Dart finding the Taung Child in a box of fossils before he was dragged off to a wedding, the origins of fossil discoveries are told and retold as vibrant oral history in paleoanthropology fieldwork-centric science. Lucy's beginning is no different.

In the mid-1970s, paleoanthropology's human origins–oriented research in East Africa was in full swing with multiple research projects exploring the potential for new fossil sites along that region's Rift Valley. Geologically, the Rift Valley is a system of volcanic rift basins that extends from the northern part of Ethiopia—or Afar Depression— running as far south as Malawi and Mozambique as the African and Indian plates continue to separate from each other. The Afar Triangle is even more geologically interesting; it is a tectonic triple junction where the African and Indian plates also separate from the Arabian one. For paleoanthropology, the Afar Triangle is an area unparalleled in the world due to the number of fossil localities exposed by the tectonic rending. Researchers like the Leakey family had worked in East Africa for decades at sites like Olduvai in Tanzania, where these fossil sites follow the geologic uplift along its fractured tectonic boundary. (The Olduvai Gorge site in northern Tanzania stretches approximately thirty kilometers along the Rift Valley; since systematic excavations began in the 1930s, it has yielded a trove of fossil hominin remains.) The Afar region was a completely new, unexplored part of the East African rift system, and its geological complexity piqued the interests of geologists and paleoanthropologists alike. As such, there were high hopes that exploration and surveys of Afar would yield new fossils that could play a role in the science of human origins.

In 1974, the International Afar Research Expedition (IARE), a geology and paleosciences research consortium, began its third formal field season at Hadar, one of these exciting unexplored sites in the Afar Depression. The 1971 IARE founding members included American paleoanthropologist Donald Johanson, French geologist and paleoanthropologist Maurice Taieb, and French anthropologist Yves Coppens, who together would build on the logical groundwork established by the Texan geologist and paleontologist Jon Kalb. (During those first years of IARE, Kalb and his family lived year-round in Ethiopia, working to keep the IARE functioning between field seasons.) Archaeologist Mary Leakey was another of the founding members, lending her reputation and expertise to the project, but she later left the consortium. By fall of 1973, the team had expanded to include other researchers and graduate students, and in November 1974 the IARE had really hit its scientific stride. Fossil locales in Hadar yielded a trove of different mammalian fossils, the team's geologic mapping progressed well, and scientific papers were published in good order. The previous season had even yielded hominin material—a knee joint that showed a species able to walk bipedally; since the fossils were around three million years old, the knee joint indicated that bipedal walking was a very old characteristic in humans' evolutionary story.

During the 1974 season, yet more hominin materials—in addition to other mammalian fossils—poured into the IRAE's collections during field surveys. The *Ethiopian Herald* publicized the team's initial discoveries that field season. On October 21, 1974, partway through that field season, the newspaper announced "Ancient Homo Sapiens Found in Central Afar" and featured a front-page photo of Ethiopian team member Ato Alemayehu Asfaw, Johanson, and a representative from the Ministry of Culture, all examining the fossil cache that included a complete upper jaw, half of another upper jaw, and half of a mandible. All of those fossils—according to that press account—were four million years old.[2]

But the real discovery of that field season came a month after the press conference—the November 1974 morning that Johanson recorded in his notes. In his coauthored 1981 bestseller *Lucy: The Beginnings of Humankind*, Johanson paints a spectacularly thrilling account of finding Lucy:

> As a paleoanthropologist . . . I am superstitious. Many of us are, because the work we do depends a great deal on luck. The fossils we study are extremely rare, and quite a few distinguished paleoanthropologists have gone a lifetime without finding a single one. I am one of the more fortunate. This was only my third year in the field at Hadar, and I had already found several. I know I am lucky, and I don't try to hide it. That is why I wrote "feel good" in my diary. When I got up that morning I felt it was one of these days when you should press your luck. One of those days when something terrific might happen . . .
>
> Throughout most of that morning, nothing did. . . . The gully in question was just over the crest of the rise where we had been working all morning. It had been thoroughly checked out at least twice before by other workers, who had

found nothing interesting. Nevertheless, conscious of the "lucky" feeling that had been with me since I woke, I decided to make that small final detour. There was virtually no bone in the gully. But as we turned to leave, I noticed something lying on the ground partway up the slope.

And it's this first-person account of Lucy's discovery that has so firmly established the genre of fossil discovery in paleoanthropological memoir:

"It's a bit of a hominid arm," I said. "That piece right next to your hand. That's hominid too."

"Jesus Christ," said Gray. "By God, you'd better believe it!" shouted Gray. "Here it is. Right here!" His voice went up into a howl. I joined him. In that 110-degree heat we began jumping up and down. With nobody to share our feeling, we hugged each other, sweaty and smelly, howling and hugging in the heat-shimmering gravel, the small brown remains of what now seemed almost certain to be parts of a single hominid skeleton lying all around us.

The camp was rocking with excitement. That first night we never went to bed at all. We talked and talked. We drank beer after beer. There was a tape recorder in the camp, and a tape of the Beatles song "Lucy in the Sky with Diamonds" went belting out into the night sky, and was played at full volume over and over out of sheer exuberance. At some point during that unforgettable evening . . . the new fossil picked up the name of Lucy, and has been so known ever since.[3]

The discovery was a partial skeleton of a very old female hominin. The petite specimen, measuring just under three feet tall, would have weighed a bit more than sixty pounds in life. In an interview with

Time magazine in 2009, Johanson acknowledged the fame and personality that are inexorably associated with the fossil. "I think she's captured the public's attention for a number of reasons," he suggested. "One, she's fairly complete. If you remove the hand bones and foot bones, she's 40% complete, so one actually gets an image of an individual, of a person. It's not just like looking at a jaw with some teeth. People can envision a little three-and-a-half foot tall female walking around. Also, I must say, her name is one that people find easy and non-threatening. People think of her as a real personality."[4]

If we measure units of knowledge in Libraries of Congress, then we measure the scientific significance and cultural importance of fossils in units of Lucy today. But how and why has she become the icon that she is?

———

After Lucy's recovery, Johanson and the Ethiopian Ministry of Culture organized another press conference, held December 20, 1974, and the next day the *Ethiopian Herald* ran the front-page headline "In Central Afar: Most Complete Remains of Man Discovered."[5] After the conference, and with the close of that third IARE field season, Lucy was whisked away to Cleveland for five years, where she was cleaned, prepped, casted, and studied. She was later returned to the National Museum of Ethiopia on January 3, 1980.

But the events that surrounded Lucy's finding were, of course, more complex than any simple narrative might imply, and the background to her origin story offers a window into the complicated relationship of science and politics. In his *Adventures in the Bone Trade*, geologist Jon Kalb offers a sobering contextualization of Lucy's discovery in the turbulent political climate of Ethiopia in 1974—or Heder 1967, in the Ethiopian calendar. While the scientific teams had been working in the Afar region, political turmoil had settled in the capital and fanned out amid the country's revolution. Kalb points out that General Mengistu Haile Mariam had executed political prisoners affiliated with Emperor Haile Selassie in the early morning of November 24, mere days before Lucy's discovery. (In his 2009 book *Lucy's Legacy: The Quest for Human Origins*, Johanson states that Lucy's discovery date was November 24, 1974, rather than November 30, 1974, as recorded in his 1981 bestseller, *Lucy: The Beginnings of Humankind*. Lucy's discovery date is generally celebrated as November 24, sharing the November date with the publication of *On the Origin of Species*.) Historian Paul Henze described that apocalyptic night: "On 23 November 1974 Mengistu sent troops.... That night 59 former imperial officials were summarily shot. All had surrendered or been arrested during the previous summer and were being held for investigation. Thus, in a single night the Ethiopian revolution turned bloody. Blood never ceased to flow for the next 17 years."[6] The juxtaposition of events—revolution and evolution—is a sobering reminder that science is a social activity, set amid its political environs. But this puts Lucy squarely in the middle of a specifically nationalistic narrative for Ethiopia.

"Later that same morning at Hadar, Johanson found 'Lucy.' . . . Thus on the day that Addis Ababa awoke to the news of the end of humanity, at least as the families of the slain understood it, the IARE celebrated the discovery of humanity's beginning," Kalb reminisces. "One of the ironies of these respective events is that a reason given for the execution of many of Ethiopia's elite was their cover-up of the famine in Wallo, where Lucy was found, and where tens of thousands of Afar nomads had died of government neglect. The two monumental events on that November day in 1974 may have marked the first time in Ethiopia's history that the Afar people and their unique land were given so much attention."[7]

Kalb makes no bones about the significance of the fossil. "Lucy was a great discovery," he asserts. "Announced at another press conference in Addis Ababa on December 20, the find was described as a

40-percent-complete skeleton of a diminutive, bipedal, adult female about one meter high. . . . The array of 63 pieces of the skeleton were found the day after Richard Leakey, his wife Meave, and Mary Leakey had visited Hadar, and there was great celebration in camp that evening." Kalb is also quick to point out that the location featured prominently in that 1974 field season. "The Lucy locality, L288, was surrounded by a cluster of seven other fossil localities mapped by Dennis Peak and myself. At one time or another in 1973, probably everyone in camp had walked across L288, Johanson included."[8] The bit of the story about the Beatles, the tape recorder, and Lucy's nickname that features so prominently in Johanson's memoirs does not even get a mention—as if it's simply too sophomoric to discuss amid the contexts of political upheavals.

———

Once it had been discovered, the next step was to formally describe what kind of fossil hominin the team had found. Lucy raised the question of what, if any, of the then recognized possible fossil species she fit into, and it quickly became apparent that she didn't line up with any known species category. Her discovery meant redrawing an evolutionary tree to take account of this newly discovered species. On March 25, 1976, Donald Johanson and Maurice Taieb published "Plio-Pleistocene Hominid Discoveries in Hadar, Ethiopia" in *Nature*. (The Plio-Pleistocene dates to roughly five million to twelve thousand years ago.) The article summarized the results of the first three IARE seasons and described the recovered remains of twelve hominid individuals from geological deposits around Hadar that were then estimated to be approximately three million years in age. Although the abstract of the paper boldly trumpeted that "the collection suggests that *Homo* and *Australopithecus* coexisted as early as 3.0 Myr ago," the crux of the issue is understated to say the least: "A partial skeleton represents the most complete hominid known from

The Ethiopian Herald

Vol. XXX — No. 1215
Annual Subscription $ 46.50
Price Per Copy 10 cents

The Press Stimulates change

MORNING NEWSPAPER
Addis Ababa — Saturday, December 21, 1974 — (Tahsas 12, 1967)

118247, 118248, 110829.
Editorial office: 112212, 111829.
P.O. Box 1074.

Advertising: 114770 117343
Sales Dept. 119050
Subscription: 111227, 115800

In Afar

Most Complete Remains Of Man Discovered

The International Afar Research Expedition announced yesterday that it recently discovered a partial skeleton of a three-million-year old hominid in the Awash Valley. This specimen is said to be representing the most complete early man discovery ever made in Africa.

The latest finding was located last November 24 by Dr. Donald C. Johanson and his student Thomas Gray at a site called Hadar. The individual was extremely small in size (about three to three and a half feet) and diagnostic features of the pelvis and sacrum have suggested to Dr. Johanson that the specimen is a female. (He already christened 'her' Lucy).

The following skeletal parts of the specimen were identified by Dr. Johanson: some hand, wrist, ankle bones and an almost complete right arm; most of the leg except for some missing fragments, a mandible with some teeth; a few skull parts, especially the back portion, ribs, parts of the backbone, and most importantly, a half pelvis with a sacrum.

For the moment, a scientific identification of this specimen's affinities has not yet been attempted.

Both Dr. Maurice Taieb, head of the French team, and Dr. Johanson, heading the American group, told the press conference held at the Ministry of Culture yesterday afternoon that the speci-

men comes from a layer of sandstone which has also produced fossils of wood, rodents, crocodiles, pigs, elephants, gazelles, some monkey teeth, and fossils of crab claws. They said the geological

(Contd. on page 5 col. 3)

Most Complete . . .

(Contd. from page 1 col. 6)

setting and the animals living with the specimen suggest that the environment was related to a beach of a vast lake that existed in the Afar region some three million years ago.

This is the third year of research for the expedition after the discovery of the site in 1968 by Dr. Taieb. Funding for the expedition has come predominantly from the National Science Foundation (United States) and the Centre National de la Recherche Scientifique (France).

The scientific direction of the expedition has been led jointly by Dr. Taieb, Charge de Recherche of the Quaternary Geological Laboratory, CNRS, France and by Dr. Johanson, Assistant Professor of Anthropology, Case Western Reserve University and Curator of Physical Anthropology, Cleveland Museum of Natural History, Cleveland, Ohio, USA. The 1974 expedition included 17 American, Ethiopian, French, and German scientists and students. This group represents a wide range of disciplines including geology, anthropology, paleontology, paleobotany, topography, etc . . .

The first fossil-man discoveries were made in 1973 when Dr. Johanson located four leg bone fragments and a skull fragment in the Hadar region. This discovery provided the oldest evidence for man's upright posture. With this encouragement the search for additional early man remains began during the 1974 field season which is just now completed.

Alemayehu Asfaw, from the Ethiopian Ministry of Culture, made the first discoveries this field season. He located a number of jaw fragments containing teeth which were announced on October 25th as assignable to the genus *Homo*. These finds are the earliest evidence in Africa for clearly recognizable human ancestors.

In total the hominids discovered by the expedition consist of parts of 10 individuals found in 11 separate localities. Dr. Taieb's geological studies suggest

that these hominids are from five stratigraphic levels representing an as yet undetermined time range. The entire Hadar sequence of sediments consists of more than 120 meters. The lowest two hominid levels are 12 meters above the base, the third is 50 meters high, the fourth 70 meters, and the fifth, containing the associated partial skeleton some 80 meters above the base.

Because of the amazing density of fossil man discoveries in only one season, and the rich associated fossils of the vertebrates living with the hominids, and the superb geological setting, the Hadar area of Ethiopia may be the most important site in eastern Africa, and therefore the world, for understanding the earliest stages of man's evolutionary past during the Plio/Pleistocene time range some 2 to 4 million years ago.

Leaders of the expedition also took the opportunity to express their appreciation to the Ethiopian Government for the opportunity to conduct research in the central Afar.

Lucy's first press conference, *Ethiopian Herald*, December 21, 1974—the first time "Lucy" appears in print.

this period."[9] This partial skeleton was AL 288-1, Lucy, and this was her first foray into the world of scientific review, much more subdued and carefully measured than the press conferences in Addis Ababa after her initial discovery. Rather dryly, the article reports:

The discovery on November 24 of a partial skeleton (AL 288-1) eroding from sand represents the most outstanding hominid specimen collected during the 1974 field season. It is obvious that this discovery provides us with a unique opportunity for reconstructing the anatomy of an early hominin in far more detail than has been previously possible. Extensive descriptive and comparative studies are projected for the AL 288 partial skeleton and will provide us with details of stature, limb, proportions, articulations and biomechanical aspects. Three weeks were devoted to intensive collecting and screening to insure the recovery of all bone fragments from the site. Laboratory preparation and analysis has only just begun, and in this report it is possible to mention only a few salient points.[10]

The language of the authors' descriptions in *Nature* is clinical—anatomical descriptions with the tone of scientific detachment we expect in academic publication. As a paper intended for scientific peers, it is filled with measurements and field methodologies. The fossil was simply discovered—it is stripped of the social commentary and detail, as its publication in a formal scientific journal demands. Here, in the context of *Nature*, the discovery was simply AL 288-1; it was a skeleton with amazing potential, but one with only four short paragraphs devoted to the fossil's anatomical descriptions and tantalizing possibilities for analysis, once the skeleton had been fully prepped.

———

"The story of how Lucy got her name is . . . more than an account of a scientific christening," science writer Roger Lewin argues in *Bones of Contention*. "It is a confection of professional and personal responses to an intellectual upheaval in the field. It is a story that, with varying degrees of clarity, reveals the swell of underlying preconceptions."[11] A fossil's name carries a lot of narrative heft. For the IARE team at Hadar, the fossil was AL 288-1. In an interview celebrating the fortieth anniversary of the fossil's discovery, Johanson reaffirmed the origin of "Lucy" as the fossil's nickname: "*Sgt. Pepper's Lonely Hearts Club Band* was the name of the album and the song called 'Lucy in the Sky with Diamonds' was playing and a member of the team suggested that we name the fossil Lucy and the name stuck."[12]

While "Lucy" came from the Beatles song, her catalog number, AL 288-1, came from the field cataloging process of that 1974 season. "AL" refers to Afar Locality and "288-1" the specific geological locality and specimen catalog number. But it's not enough for a fossil to simply have a specimen number or a nickname. A fossil needs a scientific appellation in order to have taxonomic and evolutionary status in the paleontology world. Assigning a fossil to a scientific species is an important step in giving the fossil a functioning evolutionary framework. A nickname and a specimen number give a fossil its cultural and methodological context, but it is the taxonomic name that really situates a specimen within its evolutionary framework. Assigning a fossil to a species—especially a new species—writes that fossil into an evolutionary story. If the fossil is a direct ancestor to *Homo sapiens*, we offer the species a more central, starring role in the drama than if the species is an evolutionary offshoot. In other words, if a newly described hominin species hinges on only a few bits of bone, those few bits of bone carry a lot of scientific weight in order to hold up the legitimacy of a fossil species.

But even scientific names become, themselves, historical markers and can point to specific scientific questions or famous discoverers. When Dart called the Taung Child *Australopithecus africanus*, for example, the name referred to the "southern ape of Africa," which flew in the face of evolutionary paleoconventions of the early twentieth century. *Eoanthropus dawsoni*, or "Dawson's dawn ape," inexorably links the fossil's discoverer, Charles Dawson, to the species. Likewise, *Homo neanderthalensis* denotes that the original discovery of the species came from the Neander Valley in Germany.

In 1978, four years after Lucy's original discovery, Donald Johanson, Tim White, and Yves Coppens published the article "A New Species of the Genus Australopithecus (Primates: Hominidae) from the Pliocene of Eastern Africa." This publication specifically created a new species, *Australopithecus afarensis*, that would explain Lucy's morphology and give her an evolutionary narrative. Although Johanson and geologist Maurice Taieb had published general descriptions of the hominin material in 1976, AL 288-1 didn't receive her scientific name until that *Kirtlandia* publication of 1978. With that publication, Lucy belonged to a species.

When Lucy was discovered, she joined a host of similarly curious fossils excavated in Tanzania, at the Laetoli site where Mary Leakey had been working for decades. Although history has made Lucy easily the most famous australopith, she is not, in fact, the type specimen for the species. (The type specimen for *Australopithecus afarensis* is actually LH-4, a fossilized adult mandible found in Tanzania at Laetoli.) Linking fossils from Ethiopia and Tanzania created a bit of a stir in the paleoanthropological community. First, it created a particularly intriguing argument that the distribution of this new species during the Plio-Pleistocene extended throughout East Africa; second, it created an implicit sociological link between Mary Leakey's discoveries and Johanson's. Regardless of what the type specimen was or where the geographical distribution of the *afarensis* species was, Lucy, rather than LH-4, became the cultural touchstone for talking about and making sense of human history in the Plio-Pleistocene.[13]

Australopithecus findings—discoveries before Lucy—made accurate reconstructions of any australopith species challenging. Whereas it was very difficult to imagine an entire living organism from one mandible or from some small bone fragment, a skeleton that was 40 percent complete—like Lucy's—offered enough skeletal shape to be able to put a body with the fossil relatively easily. With parts of arms, legs, ribs, and crania, not to mention the left part of the pelvis, fragments of the jaws, teeth, and several vertebrae, it was easy to match fossil to body part. Not only did the different skeletal elements provide a very visual framework for the discovery, but the presence of so many previously undiscovered skeletal parts from the Plio-Pleistocene meant that, suddenly, researchers could ask questions about hominins' niches in their environment. Since the fossil had arms and legs, it was possible to ask and answer questions of hominin locomotion— how these early hominin species would have moved. The recovered pelvis parts meant scientists could ask questions about sexual dimorphism in early human ancestors. With teeth and mandibles, paleoanthropologists and paleoecologists posed questions about the diet of *Australopithecus afarensis* and how the species would have been able to successfully consume resources from the environment.

When Lucy was announced to the scientific world as *Australopithecus afarensis*, she was the first new hominin species to be designated in fourteen years. Taxonomically, evolutionarily, and historically, the name carried a lot of clout. *Australopithecus* not only tied Lucy to Africa—similarly to how the Taung Child was tied to Africa through the name—but also established an evolutionary relationship with other fossil species. This new *Australopithecus* had to be ancestral to the genus *Homo* and related to the Taung species—this showed an evolutionary relationship between the fossil species. Even the species name *afarensis* carried some cultural provenience, tying the specimen to the Afar region where she was first discovered.

But Lucy has another moniker—her Ethiopian name, Dinkinesh. In *Lucy's Legacy*, however, Johanson describes an exchange with his Ethiopian colleague at the Ministry of Culture, Bekele Negussie, in 1974. Negussie suggested that the fossil needed an Ethiopian name and proposed "Dinkinesh," which roughly translates to "You are marvelous" from the Amharic language. "Dinkinesh" is Amharic, not Afar, creating an Ethiopian identity more broadly defined than even the regionalism implied by her scientific name, *afarensis*. In the original press conference about the 1974 field season, Lucy was introduced to the readers of the *Ethiopian Herald* as "Lucy," not as Dinkinesh, and her Amharic nickname entered the oral and written history only in more recent years. In *Lucy's Legacy*, Johanson offers the specimen another nickname in the Afar language—Heelomali—which he translates as "She is special." But for most of Lucy's life, she is just Lucy.

While *Australopithecus afarensis* electrified the scientific world, it was AL 288-1's nickname and the Beatles story that locked her into the public's consciousness. (A name, a story, were key, and certainly media conferences and television appearances did not hurt her celebrity development, either.) Nicknaming a fossil wasn't unique to

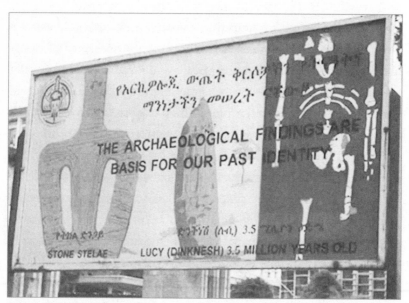

Billboard in downtown Addis Ababa, Ethiopia, illustrating the iconic role that Lucy (Dinkinesh) has played in defining national history. *(L. Pyne)*

Lucy—the Old Man and the Taung Child are a testament to the phenomena—but in Lucy's case a nickname was and still is particularly powerful. In Ethiopia, Lucy has lent her name and her cultural cachet to "numerous coffee shops, a rock band, a typing school, a fruit juice bar, and a political magazine. There is even an annual Lucy Cup soccer competition in Addis Ababa," noted Johanson in *Lucy's Legacy*, whose publication in 2009 coincided with Lucy's U.S. tour. Lucy has become a mascot-like symbol for Ethiopia writ large—giving voice and identity for enduring antiquity of Ethiopian history and prehistory.

For a fossil, names are everything: context, intrigue, history, cultural shorthand, science. The story of Lucy and her names is that of a cultural palimpsest—she is named, renamed, shaped, and reshaped by her various contexts. Her name—what she is called in what contexts and by whom—denotes the mores and complexities that surround her status as an icon.

And through her names, and their implied cachets and legacies, is how and why she has become the measuring stick by which all other fossils are compared. After-Lucy fossils are put in their "place" in scientific and popular cultures by back-sighting to that fossil. Fossils are older than Lucy. Fossils are younger than Lucy. A fossil species climbed trees—more or less so than Lucy. As a fossil, Lucy, then, is a suspended moment of evolutionary history where biology, psychology, and the fossil's cultural significance balance as a triple point of meaning.

———

If a picture is worth a thousand words, it would be hard to find a better example of an image imbued with meaning than a fossil's "official" portrait. Art historian Richard Brilliant argued that the fact that a portrait describes a real person infuses the image with more authenticity than a different kind of picture. "The very fact of the portrait's allusion to an individual human being, actually existing outside the work, defines the function of the artwork in the world and constitutes that cause of its coming into being," Brilliant wrote.[14] To that end, fossils have several types of portraits and these pictures offer another strata of authenticity to their viewers—some representations are three-dimensional and diorama-like, while others are portraits or still-life depictions of the fossil.

The 1976 *Nature* publication that introduced Lucy as a fossil full of anatomical measurements also introduced Lucy's iconic portrait. The article is accompanied by an image of the fossil carefully laid out, flat, in anatomical orientation, against a black backdrop. The figure is simply labeled "Partial skeleton (AL 288-1) from Hadar" and the portrait was reproduced again in the 1978 *Kirtlandia* article describing *Australopithecus afarensis*. Readers of the *Nature* article are left with a Rorschach-like inkblot—an image to read and let them interpret their own ideas about what the skeleton could be. But the image is really a piece of art in addition to being a figure in the journal article. This photograph quickly became one of the most iconic representations of the fossil. It's almost as if the portrait is a panel in a triptych, with Lucy's white bones illuminated and venerated against that black backdrop.

In later decades, other fossil discoveries would invoke AL 288-1 in photographs in which the white fossil bones are carefully laid out in similar anatomical position against a black backdrop. In 2009, for example, *Science* published "A New Kind of Ancestor: *Ardipithecus* Unveiled," which was the first description of the 1994 fossil find *Ardipithecus ramidus*. The striking image of *Ardipithecus* that graced the cover of *Science* evoked the iconic images of AL 288-1: *Ardipithecus* laid out in anatomical array, photographed flatly against a black background. Just about every newly discovered fossil hominin species from *Homo floresiensis* to *Australopithecus sediba* to *Homo naledi* consciously or unconsciously invokes Lucy's iconic portrait when scientists photograph their specimens in this way, appealing to Lucy's established scientific and cultural legitimacy.

In the world of three-dimensional images, though, fossil expert and renowned contemporary paleoartist John Gurche describes the process of creating a reconstruction, particularly for a specimen that has as much social and cultural importance as Lucy. Giving Lucy a physical, tangible body means that an artist instills a form and offers a life force to a fossil—it's a process that takes something static and creates a dynamic entity capable of moving, acting, and thinking. When audiences look at Gurche's reconstructions, they might not realize that they are in fact seeing hundreds—thousands, really—of artistic decisions based on decades of exacting scientific research. Over three million years ago, Lucy's appearance would have depended on her evolutionary story; in the twenty-first century, her form is a balance of artistic and scientific deductions. Providing the fossils' audiences with a face—a sculpture, a portrait, a reconstruction—creates a narrative for the fossil. The photograph, sculpture, reconstruction, or diorama freezes that narrative and invites the viewer to step into the fossils' life and to read its life story. Some of these reconstructions and visual images become culturally coded into the intellectual and public milieu and serve as important signifiers for cultural space. These pictures, these captured poses, are thus part of the public afterlife of fossils.

"As Lucy's body took shape under my fingers, it became evident that it would not be like that of any creature alive today. There are both apelike and familiarly humanlike aspects of her anatomy, but her body is not identical to either," Gurche mused, describing his work with Lucy. "The implication of the anatomical work is that, when reconstructing Lucy, she must be met on her own terms. I could not build a diminutive human form over this unique skeleton, nor could

I build that of an ape. The process reveals a body unique to her kind. Strong. Capable. And a bit wary. This is how the figure of Lucy looked to my eyes when she was completed. She is climbing down from a tree and is just dropping into an upright pose. But she doesn't do this casually. The ground is a dangerous place."[15]

The reconstructed bodies and subsequent narratives of these ancestors have been stripped bare, back to the fossils' casts. Instead of full diorama scenes, visitors see amazingly lifelike reconstructions of hominin faces, created by Gurche specifically for the Smithsonian. (Gurche's work is brilliant; his hominin sculptures make most other attempts to put a body on a human ancestor look like a bedraggled extra from *2001: A Space Odyssey*.) Gurche's reconstructions stand alone, offering an element of singularity and disconnect as the hominins stand as individuals, devoid of a scene and devoid of a story.

Earlier reconstructions of australopithecines, particularly mid-twentieth-century dioramas that featured the Taung Child or other South African australopiths, catch a lot of flak from the scientific communities today for their outdated science and outmoded presentation—ideas or hypotheses that are no longer supported by the scientific establishment. On the surface, it's easy to dismiss a diorama scene. We can say that our scientific understanding of toolmaking, social dynamics, and paleoenvironments has changed so much that we ought to dismiss these dioramas as vestiges of older, outdated science. It's easy to argue that these dioramas are doing a disservice to museum-goers since the visitors will take away "wrong" information. It's easy to take issue with the presentation of the reconstructions, saying that because the diorama stories are imprecise, it would be better to strip the scenes from the museum and display only fossil casts and their descriptions.[16]

These stories, however, humanize the australopithecines, and that's a powerful thing. It makes the fossil record accessible to us as people, not just as scientists. It makes us more sympathetic, more empathetic, with fossils we're seeing. Just as we're ready to look to the brandishing club as a clear cultural motif—à la Kubrick's *2001*—we're prepared to allow human ancestors narratives that we wouldn't have in other circumstances. Putting the body on these fossils speaks to the way that we consciously or unconsciously make sense of these scenes and human evolution more generally. Thanks to her renown—her name-brand recognition, if you will—Lucy is an invaluable character in museum storyboards of human evolution, particularly in museums like the Smithsonian. At the Smithsonian's Hall of Human Origins, a three-dimensional reconstruction of Lucy greets visitors to the human evolution exhibit. She offers museumgoers a Virgil-like presence to their tour, guiding visitors through their own evolutionary story, similarly to how Sue the *T. rex* guides visitors through the Jurassic era at Chicago's Field Museum. References to Lucy pop up in the explanatory text of the Hall's exhibits, and countless other science and natural history museums utilize Lucy as a familiar character to guide visitors through humanity's evolutionary narrative.

But the real Lucy isn't on display in any museum, even in her home museum in Addis Ababa. Visitors to the National Museum of Ethiopia see casts of Lucy's bones, while casts of other Ethiopian fossil hominin discoveries sit in glass cabinets nearby. Lucy's real bones are carefully locked away in a laboratory vault a few buildings over. Other cultural, religious, and historical artifacts displayed in the National Museum demonstrate the deep, connecting ties between past and present, giving Lucy a series of contexts. So when Lucy—her real bones—came to the United States on a touring exhibit, that tour changed up the dynamic for "Lucy in museum exhibit." Suddenly, audiences weren't simply "learning about evolution" or "science" through replicas of famous fossils; museumgoers were queuing up to view a famous object—an icon.

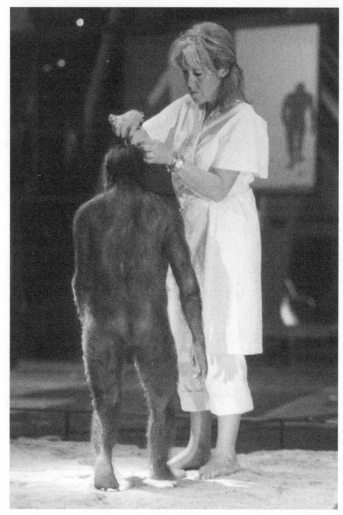

Reconstructing australopithecine Lucy. French sculptor Elisabeth Daynes of the Daynes Studio, Paris, working on a reconstruction of the Lucy specimen *Australopithecus afarensis*. Lucy is shown walking on fossil footprints that were discovered in 1976 in Laetoli, Tanzania. *(P. Plailly/E. Daynes/Science Photo Library)*

In 2007, the Houston Museum of Natural Science, in collaboration with the Ethiopian government and the U.S. State Department, launched a six-year museum tour of Lucy. The goal of the tour was simple. The hope was that a tour of the country's best-known objet d'art, the iconic hominin, would raise awareness about Ethiopian culture and give the country an opportunity to show off its heritage. "It will put Ethiopia on the map as the cradle of mankind and of civilization," announced Mohamoud Dirir, Ethiopia's then minister of culture and tourism in 2006. The same way viewers might line up to view treasures from King Tut's tomb or artifacts from Machu Picchu, an exhibit that showcased Lucy's actual bones contextualized in broader Ethiopian history, transforming the museum experience into something truly exotic—it rarefied and highlighted how special such an unprecedented loan of fossil materials truly was. Science writer Ann Gibbons noted, "Ethiopian officials had high hopes that Lucy will do for Ethiopia what King Tut's riches did for Egypt."[17]

The possibility of displaying the actual Lucy churned up a huge amount of controversy. Many were uneasy about letting such a famous, irreplaceable object leave its home country as part of a museum tour. Many museums—like the Smithsonian Institution, the American Museum of Natural History, and even Lucy's earlier home in the Cleveland Museum of Natural History—declined to host the exhibit citing concerns about possible damage to the bones. Kenyan paleoanthropologist and prominent activist Richard Leakey also objected to the transportation of an original fossil hominin outside of its country of origin. This, he and others claimed, violated a 1998 resolution established by the UNESCO-affiliated International Association for the Study of Human Paleontology that discourages the removal of fossil hominins from their place of discovery and emphasizes the use of replicas in museum displays. "If we start sending these fossils out of the country, Kenya and Ethiopia cease to be places where you can study fossils. It immediately changes the role of the museum as a place for scientific study," said Leakey, former director of the National Museums of Kenya.[18]

Dirir countered Leakey's position by arguing that Ethiopian officials thought that spreading the word about Lucy and their nation's rich cultural heritage could help draw tourists to Ethiopia and change its image. "The money will go to museums, and just to museums," said Dirir. "Just keeping fossils in Ethiopia will not develop science, museums, or the custodians of these fossils." (In addition to building awareness about Ethiopian history and human evolution, Lucy's trip to the United States would provide new data from the fossil. While in Texas, she was CT-scanned at the University of Texas, Austin—the scans [data] generated were transferred to the National Museum in Addis Ababa for future researchers.) Prominent Ethiopian paleoanthropologist Dr. Zeresenay Alemseged, then of the Max Planck Institute for Evolutionary Anthropology in Leipzig, Germany, and 2001 discoverer of Selam (dubbed "Lucy's Baby"), was skeptically unimpressed. "What Ethiopians are benefiting from this? I have not seen a document that clearly defines the role for the National Museum of Ethiopia," he said in an interview with *Nature* in 2006. "I have never heard of any Ethiopian paleoanthropologist being involved. If money is being generated, it should be clear what percentage will go to Ethiopian science."[19]

Once Lucy was returned to Ethiopia in 1980, via Cleveland, she went to the National Museum in Addis Ababa, kept under close lock and key. In the midst of newly emerging Ethiopian nationalism, a fossil of such iconic status was a particularly powerful cultural symbol about the longevity of Ethiopian history via its own prehistory. Researchers interested in working with any kind of Ethiopian fossil or archaeological materials were required to travel to the National Museum for their studies; scientists who wanted to study Lucy needed to have the approval of the museum and fossil keepers and then would come study, caliper, and measure Lucy on her own turf, in a sort of data-driven pilgrimage.

Fast-forward almost thirty-five years. Texan congressman Mickey

Leland had developed a strong rapport with various ministers in Ethiopia through his work with food and humanitarian agencies, working to get more financial resources into the country. Having traveled extensively there throughout the 1980s, Congressman Leland had a well-established reputation within the Ethiopian community. After his death in a tragic plane crash (while on a mission to alleviate the severe famine that affected Ethiopia in 1989), consul generals from Ethiopia wanted to find a way of commemorating his work and legacy, even decades later. They offered a tour of their most well-known icon—a tour of the fossil that hadn't left the museum after it was originally returned to the institution in the early 1980s and the fossil that was, perhaps, the most well known within the popular and scientific communities. Before the tour began, in 2007, there were years worth of negotiations and logistics to be addressed.[20]

In 2003, Dr. Dirk Van Tuerenhout received an unexpected phone call. As curator of anthropology for the Houston Museum of Natural Science, he was accustomed to fielding questions about the market potential for particular exhibits, coordinating the display of rare and privately owned artifacts with odd bequests and requests within the museum. Throughout his career, Van Tuerenhout has helped curate an incredible variety of exhibits—including *The Dead Sea Scrolls* (2004), *Mummy: The Inside Story* (2005), *Secrets of the Silk Road* (2010), *The Cave Paintings of Lascaux* (2013–2014), and *Magna Carta* (2014). When Van Tuerenhout picked up the phone one day over lunch, he was sure that he was going to field some relatively wingnut inquiry about some recent exhibit that the museum had run. Instead, he was floored by the conversation.[21]

The woman on the other end of the phone introduced herself as from the Texas office of tourism. The tourism official asked if the Houston Museum of Natural Science organized any archaeology exhibits, and if so, did people come from far enough away to require a hotel stay. Van Tuerenhout politely answered affirmatively to both questions. He was, however, a bit stymied by the question of "How many people traveled to the museum and then stayed the night at a hotel in Houston?" He said he had no idea and then, in turn, asked what all of these questions were about. The tourism official replied that there was the possibility of an exhibit about Ethiopia coming to Houston, Texas, and that "Ms. Lucy was going to be part of it."

It would be hard to find an anthropologist worth his salt that wouldn't make the connection between "Ms. Lucy" and the iconic fossil hominin, and Van Tuerenhout chuckles at the memory. That conversation was the beginning of a long partnership between the Houston Museum of Natural Science and the National Museum of Ethiopia as Van Tuerenhout and others worked to organize the display of Lucy and other Ethiopian artifacts.[22]

Negotiating Lucy's twenty-first-century tour was no small task. Curators, scientists, art historians, politicians, and various Ethiopian groups all had different, sometimes vying, interests in seeing whether the exhibit would come to fruition and if so what shape it might take. Meetings between American and Ethiopian museums to negotiate the exhibit and which artifacts would travel with Lucy went on for months. Even before the question of which religious artifacts, triptychs, or other icons would be allowed to leave the National Museum, serious questions were raised by the Ethiopian Orthodox bishops about how artifacts would be treated while on loan and about their safe return. More than triptychs or processional crosses, however, most of the discussion about loaning artifacts centered on the safe return of Lucy and concerns about damage to or loss of the fossil. Losing Lucy would mean losing a crucial part of Ethiopia's modern history and prehistory.

In many ways, it would seem that their concerns hinged, however, on understanding what kind of object Lucy was—or is. Defining the *kind* of object Lucy was would translate to how museums and audiences ought to view her, but it also meant that different audiences contributed different types of expertise in creating the "Lucy" that would be on exhibit—understanding how a scientific object like a fossil can draw from the social cachet that something like a religious icon carries. And certainly, within the Orthodox Coptic tradition in Ethiopia, icons have a specific place and role. Painted in bold, simple colors with heavily accented eyes, icons serve as religious reminders, sure. They also serve as cultural testimonies—an affirmation of narrative and cultural place. Characterized by biblical scenes that show classic tropes of religious iconography—the Annunciation, the Nativity, the Crucifixion, the Ascension—the icons are often painted as parts of a series or as triptychs. In addition to scenes from the life of Christ, we see other common motifs, like Saint George and the dragon. The paintings are characterized by removing the person (or the scene) from any kind of specific environment or context.[23]

However, this goes beyond a simple observation that religious tropes permeate specific cultural contexts. Thinking about Lucy in terms of her association with Coptic iconography deepens her cultural role and the explanatory power she wields. As an icon herself, she becomes a character acting out a moral universe and a moral philosophy as part of Ethiopia's national story. Her story is one of a fossil icon's ascension, and this icon—"Dinkinesh" in Amharic or "Lucy" in the popular vernacular—translates well beyond its simple scientific context.

Art and curatorial worlds have asked and answered the question of an artifact's objectness over and over; these objects are insured and shipped around the world for exhibits and tours. Unlike natural history exhibits that show casts of fossil bones, no art museum is going to advertise showing a copy of a Picasso or a faux Matisse. There's an implied authenticity about the objects in the art world on display in museums that doesn't seem to necessarily hold true for paleoanthropology exhibits. If the objection was that Lucy was a rare object, there are means and methods for shipping and displaying rare and irreplaceable items. But all objects—overtly scientific or not—act as cultural signs and symbols that tap into our senses and transfer information from objects to us, communicating the intended meaning wrapped up in those objects.

The museums that did opt to host the Lucy and Ethiopian exhibit—like the Pacific Science Center in Seattle or the Houston Museum of Natural Science—were able to offer their audiences what Van Tuerenhout argued was a completely unique and important educational opportunity. (In an interview with the *New York Times,* Joel Bartsch, the president of the Houston Museum, estimated that in the year that Lucy was on exhibit in Houston, she drew approximately 210,000 visitors. That's a huge number of museumgoers checking out Ethiopian history, prehistory, and fossil history.)[24] Casts, reconstructions, and images of famous fossils help us to become aware of scientific conversations and the "doing" of paleoanthropology, to say nothing of the power of authenticity. Anthropologist Kristi Lewton

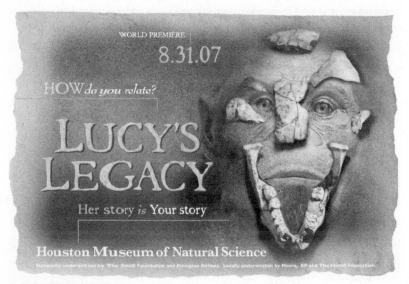

Media advertising for *Lucy's Legacy* exhibit, 2007. *(Image courtesy of Houston Museum of Natural Science)*

Media advertising for *Lucy's Legacy* exhibit, 2007. *(Image courtesy of Houston Museum of Natural Science)*

and I had the same reaction upon seeing the Taung Child in person; people have a never-ending desire to figure out who "really" perpetrated the Piltdown Hoax, or to discover the "real" Peking Man fossils. These impressions—replicas, photographs, pieces of scientific ephemera—are not, however, the same as seeing The Real Fossil.

Every exhibit is made up of hundreds—thousands—of choices, both small and large. Choices about what to exhibit where, and how, with these choices leading to other decisions about how objects ought to be transported, stored, and curated before, during, and after an exhibit—*Lucy's Legacy* was no different.

The curatorial team included Dr. Nancy Odegaard, the head of preservation at the Arizona State Museum in Tucson, as well as Ronald Harvey and Dr. Vicki Cassman. This team orchestrated the incredible logistics that surrounded Lucy's exhibit. "As a conservator, I find myself as an advocate for the object," Odegaard commented. "Once the decision is made that an object will travel, I work through the potential problems and balance the risks. But first and foremost, I am an advocate for the objects, since they cannot speak for themselves."[25] For the thousands of visitors who saw Lucy over her six-year exhibit, what they saw, in reality, was the result of thousands of decisions that the curatorial team had made long before Lucy ever left Ethiopia.

A paleoanthropologist might read a fossil species' evolutionary history through a fossil, but a curator can read a cultural history of that fossil in the way that the fossil has been treated postexcavation. These curators are not "artifact technicians," but experts whose knowledge means that the fossils can be studied by scientists and seen in museums. They know what type of glue can hold a fossil together and which will cause yellowing and deterioration. They know how to inventory and store artifacts appropriately. "We can see changes on bone and fossil," Odegaard observed. "As a conservator, I can see a cultural history written on the fossil, where glue might be discolored or a museum number has worn off. Over time, even measuring a fossil with calipers will cause wear and tear on the bone."[26]

Odegaard, Cassman, and Harvey traveled to Ethiopia to see Lucy in the National Museum so they would have a better sense of what would be required to move her. The curators inventoried the fossil's skeleton and took stock of how she was then cataloged in the museum. They built the travel case, using casts of Lucy to see how everything would fit together and how it would be taken out, going through customs and at the different museums. "Lucy rode in the first-class section of the plane, in the overhead luggage compartment, with her seventy-six fossil elements divided between two Pelican cases that would have floated if, god forbid, the plane crashed in the ocean," Ron Harvey said.[27]

Odegaard designed and constructed little custom plastic ziplock baggies for each of Lucy's bone fragments, so that no one would have to touch the fossil itself if The Case was opened for a customs inspection. ("It was definitely 'The Case,'" Odegaard laughed. "The Case, with capital letters.") Every bone piece had a photo of the front and back of it attached to its corresponding baggie. "This system meant that it was immediately obvious if something was missing or if something was amiss," Odegaard recalled. "The only person who would actually touch the fossil was Alemu Admassu, a curator at the National Museum. The only people in the room with conservator Ron Harvey when she was being boxed and unboxed were Alemu and the director of the museum. Limiting the number of people in contact with the fossil was a way of limiting the potential for damage."[28] (In July 2015, President Barack Obama met Lucy during his tour of East Africa, which included a stop in Ethiopia. Both President Obama and the fossil received a motorcade through the city. Before a state dinner at Ethiopia's National Palace, Ethiopian paleoanthropologist Dr. Zeresenay Alemseged gave an informal, impromptu demonstration of Lucy's anatomy and encouraged President Obama to touch the fossil. When some colleagues questioned this, Alemseged was quoted in the *Washington Post* as saying, "Extraordinary people have extraordinary access.")[29]

Odegaard, Harvey, and Cassman went through several dry runs with

the case, loading and unloading replica casts, trying to find any lingering issues with their transportation system, and seeking ways to reduce
the potential for damage to the fossil. Harvey took before-and-after
photos of Lucy at every museum stop, and those photos were used to
assess the fossil when she was returned to Ethiopia. "There was no damage at all during her tour," Odegaard noted. "It felt unprecedented."[30]

In March 2007, I saw Lucy in her *Legacy* exhibit in Houston and
it was unlike any paleoexhibit that I had ever seen. Most traditional
fossil exhibits are large, brightly lit halls, places where families are
crammed around dioramas and reconstructions, where docents shout
to make themselves heard over jubilant school groups. The ambience
surrounding Lucy was very different—the room with her bones had
a different vibe from the rest of the museum and even a different feeling from the adjacent brightly lit rooms filled with Ethiopian artifacts.
Where other parts of the museum were loud and boisterous, the darkened room where Lucy lay felt very subdued and reverent. The public
filed past the fossil much as if attending a wake, with Lucy laid out
prone on a table in anatomical configuration. Perhaps a more apt allusion would be that the visitors were filing past a religious relic—the
darkened room, the solemnity of the display, and the unlifelike pose
of the fossil created an atmosphere that suggested Lucy's life was different from the articulated mammoths and dinosaurs two exhibit halls
over, which were fossil replica skeletons arranged midmotion. The
exhibit highlighted the complexities of an iconic celebrity fossil and
the difficulties of moving her between audiences. The tour was an
intricate pilgrimage for both the public and the fossil; countless scientists, paleoenthusiasts, and museumgoers trekked to hosting museums
for the once-in-a-lifetime opportunity to see the fossil.

In other museums of science, where replica casts of Lucy are traditionally displayed in motion against painted environmental back-
drops, the solemn setting in the Houston museum resembled a
medieval reliquary for hominin ancestry. Here, Lucy intermixed with

Coptic Christian crosses and paintings (plus stone tools and paleo-magnetism rock samples from East Africa's Rift Valley) as a latter-day icon—a mixture of science, culture, and emotion.

"As a scientist, I am treading dangerous ground by speaking of Lucy in 'mythic' terms. Lucy's real significance, of course, is not to be found in symbol; it rests in whatever empirical evidence she provides for the understanding of the process of evolution, specifically, the evolutionary origins of the human species," Donald Johanson offered in his 1990 bestseller, *Lucy's Child*.[31] But according to nineteenth-century linguist Ferdinand de Saussure, signs and symbols signify cultural intent and meaning. We read these societal clues almost as involuntarily as we breathe; they're a language of making sense of our surrounding material culture, and it's practically impossible not to draw from our societal clues to interpret what we're internalizing.[32]

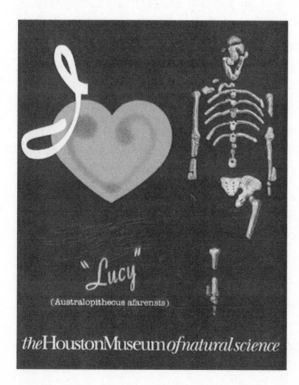

Houston Museum of Natural Science gift store swag from *Lucy's Legacy* exhibit, 2007. (*L. Pyne. Image courtesy of Houston Museum of Natural Science*)

———

Ever since her discovery, just about all the media coverage of any other fossil discovery uses Lucy as a comparative point. New fossils are ancestors that Lucy shares with us or not. A fossil like Selam—a three-year-old juvenile australopithecine discovered by Zeresenay Alemseged in 2001—has been dubbed "Lucy's Baby," lending a cultural familiarity to Selam's relatively recent fossil discovery (published in 2006). Popular science books like *Lucy Long Ago, Lucy's Child, From Lucy to Language,* and *Lucy's Legacy* invoke power (aside from alliteration) in their appeal to their audience understanding and knowing Lucy as a cultural sign and symbol.

Conservator Ron Harvey spent six years as part of Lucy's entourage as she traveled from one museum to another. "She really is the grande dame of science," he reflected. "I think she connects with humanity on a level that I haven't seen with any other artifact."[33] Kristi Lewton pointed out to me that Lucy is a useful way to explain her research to nonspecialists, and she knew several other anthropologists who did the same. We say to people, "We study the evolution of hands, feet, locomotion, pelvis, etc. You've heard of Lucy? Well, Lucy did or had x and we, as *Homo sapiens*, have y."[34]

The story—the iconography—of Lucy has coalesced and is told and retold. The prominent casting company Bone Clones noted that its sales of Lucy casts have held steady over the last decade, due in no small part to her draw in an educational context. "Lucy had become iconic with the public and so schools wanted to teach anthropology using it as an example. The public has known and thought of Lucy as the 'mother' of all modern humans. Of course, it is an exaggeration but still a great teaching tool."[35]

Could there be another Lucy? Could there be a fossil that carries the same cachet that she does? Yes and no. Yes, as more and more fossils are found and enter the cultural vernacular from their scientific publications, it's possible that another fossil could come to have the same nationalistic, scientific, and iconic status that Lucy currently enjoys. Lucy—like all objects—is a product of her various contexts. Part of what propelled her to such iconic levels were the times and places that contribute to her life and afterlife; these various contexts simply can't be manufactured, and the time it takes for the fossils to come to life simply can't be shortchanged. There won't—there can't—be another Lucy. But that can't and won't stop other fossils from trying.

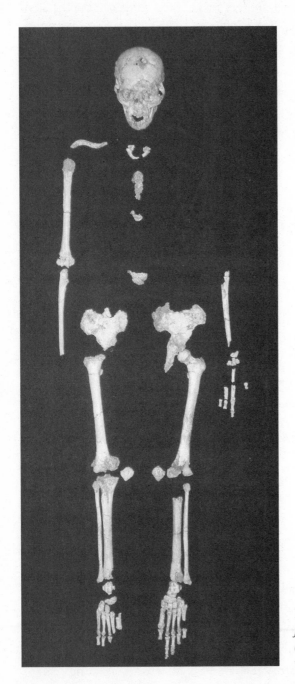

Portrait of LB1, *Homo
floresiensis.*
*(William Jungers. Used
with permission)*

CHAPTER SIX

THE PRECIOUS: FLO'S LIFE
AS A HOBBIT

I n the early morning we went to the site, and when we arrived in the cave, I didn't say a thing because both my mind and heart couldn't handle this incredible moment," archaeologist Dr. Thomas Sutikna described, recalling the 2003 field season at Liang Bua on the island of Flores, Indonesia.[1] After months of excavations and years in planning, the team had finally made a fantastic fossil discovery—one that completely shocked the world of paleoanthropology. They had unearthed a truly amazing humanlike fossil, a bizarre specimen that immediately piqued a plethora of scientific disputes that have lasted over a decade. Although the small, three-foot-tall adult hominin was named *Homo floresiensis* in scientific literature and christened "Flo" by the research teams, she is best known to the rest of the world as the real-life hominin hobbit.

By that 2003 Liang Bua field season, it had been well over ten years since any new fossil hominin species had been described in scientific literature. While fossils continued to enter the scientific record, they were easily assigned to well-established fossil hominin taxa, like Lucy's *Australopithecus afarensis* or the Taung Child's

Australopithecus africanus. So the discovery of Flo and eight other small hominins on Flores was fundamentally a game changer in human evolution studies. Finding a hominin so small, with such a small brain, so relatively late in the geologic record, and in Southeast Asia shook up how scientists thought about human evolution. This fossil challenged the who, what, when, and where of hominin evolutionary history—in other words, the discovery was something new, it was unexpected, and it was utterly confounding.

As important as the discovery was within the scientific community, the *Homo floresiensis* discovery amassed an enormous amount of public interest as the announcement of the fossil neatly overlapped with the final installment of the 2001–2003 film release of the *Lord of the Rings* trilogy. Thanks in no small part to Elijah Wood's prosthetic-eared portrayal of the hobbit Frodo Baggins, by the time the Flores fossils were excavated and published, the world was primed to think big about small creatures. A cute little hominin was exactly what was on the world's mind when the fossils were discovered—and a cute little hominin was what everyone got. When *Nature* published the Flores specimens in October 2004, I was a grad student spending a semester working on a paleoarchaeology project in coastal South Africa. The discovery shocked everyone on the project; the field director kept saying, "It was *this* big!!! *This* tall!!! This is crazy! It really is a hobbit!" He couldn't stop gesturing about the species' diminutive size—indicating the small stature with his hand at his waist. "What's next? Gandalf? A Legolas? Should we ask the NSF to fund excavations at Mordor?!?"

Once a fossil is discovered, it is assigned to a species—either a new one or one that is already established. For many famous discoveries, like Lucy, the Taung Child, or even Piltdown, the bones were so different from what had been found before that the fossils justified creating that new species. All fossil discoveries are then fitted into phylogenetic trees—scientists describe where these fossils fit in terms of their evolutionary history, giving them an evolutionary narrative and context. (Is this species ancestral to another? Is it more or less like this other species?) But new fossil discoveries must also fit into a cultural context. For some—like the Taung Child—the struggle to fit into hominin evolution *is* its cultural story. For others—like Piltdown—the fit of the fossil to a cultural narrative is prefabricated, and the cultural story is the dismantling of that forced pairing of a fossil so perfectly designed to fit an evolutionary paradigm. Flo's story, however, is very different from other celebrity fossils and offers a new model for a celebrity fossil. Flo simply didn't fit the then existing models of phylogeny—she was too small and too recent in the geological record for her to easily slide onto a branch of the hominin family tree, and her cultural story practically mapped onto a preexisting template, already at hand with *Lord of the Rings*. It's as though the fossil was discovered to give meaning to a flourishing cultural meme—she was a celebrity before she was discovered, and this cultural embodiment is what makes her fame so unique.

———

Back in 1995, archaeologist Dr. Mike Morwood was a lecturer at the University of New England in New South Wales, Australia. For years, his research emphasized Australian Aboriginal archaeology, particularly in Kimberley, northern Australia, in areas that would be likely beachheads for the first people to reach Australia's shores from Asia during the Pleistocene, sometime between 2.6 million and 11,000 years ago. (Although most current research puts the period of these first Australian migrations between 40,000 and 60,000 years ago.) After spending years researching the Australian side of early *Homo sapiens'* migration path, Morwood found it impossible to not wonder how and from where those first migrants from Asia came, since all of the possible routes involved crossing the biogeographic boundary known as the Wallace Line—an invisible demarcation that describes the separation and pattern of how plant and animals species separate from mainland Asia, the Asian islands, and Australia. Pleistocene migration to Australia could have followed several equally possible routes: some migrations could have followed prevailing ocean currents, with landfalls at the western tip of New Guinea and the island of Aru (part of the Greater Australian coastline during the Pleistocene); others would have involved island-hopping from the Nusa Tenggara islands from Lombok to Flores to Kimberley via Timor. Morwood's interest in moving his research's emphasis from Australia to Southeast Asia, specifically Indonesia, was a way of stepping back research into the question of the first Australians, tackling finally what he saw as "big questions" in archaeology and paleoanthropology.

By the mid-1990s, Morwood began writing to Indonesian researchers to explore the possibility of a collaborative project that would examine these three potential migration paths in the Wallace Line area for early *Homo* species. Frustrated by the slow pace of development, eventually Morwood simply went to Jakarta and introduced himself to Professor Raden Pandji Soejono at the National Research Centre for Archaeology (ARKENAS) and met with Dr. Fachroel Aziz, a paleontologist with the Geological Research and Development Centre (GRDC) in Bandung, two men interested in his project proposal. Aziz was immediately enthusiastic about a joint project because he and his team had been finding stone tool artifacts at various archaeological sites for years. A few small-scale projects around Flores provided the basis for grants that would allow a joint Australian-Indonesian team to excavate. This joint project became the foundation for the work that would eventually lead to the 2003 field season.

The project slowly expanded to include other archaeologists from GRDC, ARKENAS, the University of Gadjah Mada, and Northeastern University in Australia, and fieldwork began in the Soa Basin in

2001. Researchers visited different caves in the basin, namely Liang Bua and Liang Galan. "Stepping into the cave for the first time, I was immediately struck by its size, and particularly impressed by its suitability for human occupation: it was spacious, well lit with a northern outlook, and had a flat, dry clay floor, which would have made it a comfortable place to live," Morwood recalled after his initial visit to Liang Bua.[2]

Excavations at Liang Bua, discovery site of *Homo floresiensis*. *(Photo courtesy of Wikimedia, CC BY-SA-2.5)*

A project at the Liang Bua site would build on earlier archaeological work conducted by Dutch missionary and amateur archaeologist Father Theodor Verhoeven in the 1950s as well as excavations by professional archaeologists such as Professor Soejono in the 1980s. In March 2001, preparations began in earnest for modern excavations at Liang Bua— excavations that would be undertaken under the authority of ARKENAS with collaborative publications specified. On April 10, 2001, Morwood and colleagues flew to Kupang in West Timor to get excavation permits from the departments of Culture, Police, and Social Politics.

"Liang Bua is a very easy site to work on," Morwood explained. "It is not until you step inside that you realize how big it is. An intrusive concrete archway and path now installed by the Manggarai provincial government leads up to a lockable gate in the high mesh and barbed wire fence that restricts access to the cave. The key is kept by the official cave guardians Rikus Bandar and his son Agus Mangga, who also act as guides for the few tourists venturing this far out of Ruteng, the provincial capital."[3] Liang Bua was used as an elementary school in earlier decades when Father Verhoeven first arrived in Flores. (Liang Bua was one of the many sites Father Verhoeven had excavated during his seventeen years as a resident priest and prehistorian on Flores.) In 1950, Verhoeven decided that the cave would make an excellent excavation project, and this neatly coincided with the opening of a more conventional schoolhouse. He excavated a small test pit in the west wall of the cave, just inside the entrance, that backed up to some roof fall in the back of the cave.

The early excavations at Liang Bua in the twenty-first century yielded an amazing assemblage of artifacts. Once researchers pried their way through the hard flowstone of the cave's floor, the subsequent layers of clay were chock-full of artifacts: stone tools, bones, and teeth—up to five thousand artifacts per cubic meter of deposit. Close to two hundred tons of the cave's sedimentary deposits were processed for artifacts each season. The presence of artifacts like these indicated the presence of very old human activity in the cave; the question was just which species. When researchers found a small piece of humanlike arm bone—a radius—at a depth of around six meters, excavation efforts redoubled. "To keep tabs on how things were going at Liang Bua, I was phoning Hotel Sindha in Tuteng every night to get a summary of progress, finds and problems," Morwood wrote. "On August 10, Thomas [Sutikna] answered the phone as if he had been sitting right on top of it. Bursting with excitement, he told me that they had just found the skeleton of a nonmodern child in

Sector VII at a depth of six meters. They had found it! They had found the hominid that went with the *Stegodon* bones and artifacts. The very first year of our project was off to a flying start."[4]

"Before Mike Morwood left for the season in 2003, I said, 'Why are you leaving now? If you leave, maybe we will find something important.' A few days later, on 2 September, I was supervising sector VII. Our local workers were digging at around 5.9 metres. Their trowel met with a skull. A member of our team who specializes in animal and human bones came down and said, 'Yes, I'm sure that's a human bone. But it's very small.'" Wahyu Saptomo, a field archaeologist, recalled that moment of discovery ten years later in an interview with *Nature* correspondent Ewen Callaway. Saptomo realized the magnitude of what the team could be looking at immediately: "Thomas, he was sick and was at the hotel that day. So I went back and met with him. I said, 'We have something very important. We found the first hominid in the Pleistocene layer.'"[5]

The discovery of the hominin bones was a hugely significant moment in the Liang Bua excavations, unlike anything discovered before at the site. Until any hominin bones were recovered, tying the extensive stone tool assemblages found at the site to a specific species, let alone trying to figure out how the artifacts were used, was practically impossible. So when an actual hominin skeleton was unearthed, researchers knew that they would be able to tie the fossil hominin to the stone tools that they had recovered in the cave's sediments. Recovering the bones during the excavations, however, was incredibly difficult as the bones were exceptionally fragile—in fact, they hadn't actually undergone the fossilization process and were, according to the research team, like "wet blotting paper" since the soils they were found in were so moist.[6] Preservation of the bones would be—and was!—problematic and researchers were frantically concerned about damage to the bones since they were so soft.

To date, remains from a total of nine individuals have been recovered from the site, including one complete skull. When the bones were first recovered, on-the-ground researchers and excavators didn't know what exactly they were looking at. Something "human" writ large, of course (or small, in this case), but whether it was an adult or a child, what species of *Homo* it could be assigned to, and how old were simply open questions. Morwood sent a sketch to his colleague paleoanthropologist Dr. Peter Brown, for him to weigh in on the specimens, inviting him to Flores to take a look at the discovery.

"Mike [Morwood] doesn't know much about human skeletons, and the Indonesian researchers didn't either. I was quite sceptical. The drawing may as well have been a Greek urn in terms of looking like anything much at all," Brown recalled in a 2014 interview with Callaway. "I was interested and willing to go to Jakarta. It's an interesting place to visit. I like the food. I like the atmosphere and the culture and everything else, but I didn't expect to find anything interesting or important. At the most, I thought it was going to be a sub-adult modern human skeleton, probably dating to the Neolithic period or maybe a little bit earlier. The other possibility was a pathological individual, someone with a growth disorder. Those were my expectations when I turned up."[7] Brown, and the rest of the scientific community, would soon discover just how mistaken they were. The skeletons (plural) turned out to be anything but banal.

———

In 2004, the Liang Bua research team officially published their discovery in *Nature*. Thanks to *Nature*'s media embargo, no whispers of the find swirled through the paleointelligentsia before publication, so the announcement sent shock waves through the scientific world. In the paper, researchers described the anatomy of LB1 ("Flo") and designated the fossil as the type specimen of the *Homo floresiensis* species—a new species, because the size and shape of the bones were so different from any other set of fossils collected. Researchers highlighted the unique characteristics of the bones in LB1's nearly complete skeleton—pronounced as an adult female who stood about three feet tall, weighed between thirty-five and sixty-five pounds, and died about eighteen thousand years ago. LB1's cranium was small—about the size of a chimpanzee's—and the shape was confounding. This Flores specimen—a small, rather hobbit-like hominin—could walk bipedally, and archaeologists found evidence that pointed to the species' controlled use of fire, spearheads, and group hunting—all unlikely complex behaviors for this new fossil species that wasn't *Homo erectus*, *Homo neanderthalensis*, or *Homo sapiens*.

The skeletal remains of the small hominins from Flores were met with shock but also scrutiny. On the ten-year anniversary of the Flores discovery, paleoanthropologist Dr. William Jungers, who had participated in many of the anatomical studies of the fossil after its discovery, recalled, "I had to check the date to make sure it wasn't April Fool's Day. It was so preposterous on the surface that there could be this little hominin that evolved in isolation in southeast Asia for God knows how long and persisted until almost the Holocene."[8] Since geologists put the beginning of the Holocene about eleven thousand years before the present, Dr. Jungers's comment underscored the fact that *Homo floresiensis* had lived until a very recent time in terms of the fossil record. The scientific community seemed split as to how to best interpret the small size of the fossil specimens—disease? a new species? genetic aberrations?—and quickly put Southeast Asia back into paleoanthropology's spotlight.

Part of introducing the specimens to the paleoanthropological community meant that the fossils had to be assigned to a specific taxa. They had to be given a scientific name and assigned to a species, and bestowing a name implicitly connotes a phylogeny and evolutionary life history. Assigning the fossils to *Homo* rather than *Australopithecus* offers a very different narrative about hominin mobility and dispersal in the Pleistocene. Assigning the species to *Homo erectus* would have had different implications for how much variation was acceptable within a single species. A completely new genus and species name would have meant that the fossils' morphology was so different that there wasn't an evolutionary narrative thread that could offer continuity between previous discoveries and the Flores discovery.

Ultimately, the team of researchers—with feedback from peer review within the scientific community—settled on *Homo floresiensis*. Henry Gee, then senior editor at *Nature*, recalled some of the difficulties that surrounded the specimen's taxonomy. "When it came to us, they had given it this Latin name, *Sundanthropus floresianus*—man from the Sunda region from Flores. Well, the referees said it's a member of *Homo* so that's what it should be, and one of the referees says *floresianus* actually means 'flowery anus' so it should be *floresiensis*. So *Homo floresiensis* came along."[9]

More than shake up the hominin family tree, the Flores fossil set the tone for twenty-first-century paleoanthropology. Its discovery created the sense that paleoanthropology still had new and curious fossils to find and these fossils could—would!—be found in unexpected places and have significant implication for how we think about evolution.

———

Historical context for Flo's story goes back much further than the 1950s and early excavations at Liang Bua. Paleoanthropology's roots in Southeast Asia trace themselves to the nineteenth century with Eugène Dubois's discovery of Java Man. Dubois's discovery kicked off over a century of archaeological and paleoanthropological research, and scientists have been working off and on throughout Indonesia for decades. While the island archipelago could boast other hominin discoveries after Dubois's initial work—for example, "Solo Man" (Ngandong) was discovered between 1931 and 1933 by German-Dutch paleontologist Gustav Heinrich Ralph von Koenigswald in Java—the Last Big Find from Southeast Asia that fundamentally shook up both paleoanthropology was the Java Man fossils, found in 1891.

When Dubois published the fossil species he'd discovered and called it *Pithecanthropus erectus*, he was quick to argue that he had found the "missing link"—an ancient ancestor that was clear evolutionary proof of humanity's antiquity. (*Pithecanthropus erectus* was later renamed *Homo erectus* in 1950 by biologist Ernst Mayr, once Mayr had examined the Java and Zhoukoudian specimens. The similarities in the skeletons led Mayr to conclude that these fossils, separated by time and space, were actually members of the same fossil species.) At the turn of the twentieth century, the impetus to find humanity's "missing link" to apelike ancestors underscored scientists' fossil-searching research agendas. These missing links were thought to be those species that showed anatomical characteristics along a continuum of apelike to humanlike—a view of evolution that we now call unilinear.

Dubois occupied a rather curious niche in the early days of paleoanthropology: he operated as a well-qualified amateur, meaning he didn't have an academic or an institutional post. However, his anatomy training and medical background—to say nothing of his extensive reading about the late nineteenth-century missing link fossil discoveries like Europe's Neanderthals—offered enough expertise for Dubois to know what he was looking for and to know when he had found it. Without institutional support or independent wealth, Dubois took a post as a medical doctor in the Dutch East Indies, knowing that the post was a means of getting to Java, a place he was convinced would yield ancient human ancestors. He began surveys in 1887, employing local islanders to search for fossils, concentrating his efforts on Trinil in East Java and Sangiran in Central Java. In 1891, a small but significant set of fossils was recovered from sediments that filled the banks of the Solo River. This assemblage—a tooth, a skullcap, and a thighbone—became the first non-Neanderthal species to enter the history of paleoanthropology.

Dubois welcomed the excitement and enthusiasm that surrounded his *Pithecanthropus* ("apelike man") discovery in 1891–1892, and he argued that the fossil remains were proof of a missing link between ape ancestors and modern-day humans. The Java Man fossils became an immediate scientific sensation and one that generated sheaves of articles and scientific papers. However, by the early twentieth century, the fossils also stirred up a great deal of controversy since many researchers were dubious about the existence of "transitional species"—"missing links"—and if these fossil species did exist, scientists were reticent to allocate these species outside of Europe. The bright lights that surrounded Java Man began to fade and the curtain inched its way down.

The lights faded so much, in fact, that the scientific fame that Dubois had enjoyed at the beginning of the twentieth century proved fickle when new fossils (like Peking Man and Piltdown) entered the paleo community and scientists suggested different hypotheses for evolutionary patterns, particularly questioning how useful notions like "transitional species" really were. In response to this criticism—which he took as personal attacks—Dubois took his *Pithecanthropus* fossils and figuratively went home, restricting scientists' access to his Southeast Asian specimens. Convinced that the scientific community was out to persecute and mock him, Dubois denied researchers access to the bones for any subsequent studies. If researchers couldn't study the bones, his logic went, then they couldn't draw conclusions that might contradict his own.

Regardless, Dubois's discovery firmly situated Southeast Asia as the place to go to find missing links; *Pithecanthropus* de-emphasized Africa and even Europe to some extent. The story of Dubois, his Java Man, and the early days of paleoanthropology serve as a historical prototype for the Flores discovery more than one hundred years later.

————

Attempts to explain Flo's anatomy offer a particularly interesting parallel with paleoresearch from the nineteenth century, beyond her historical associations with Java Man. In fact, the ways that scientists have attempted to explain why Flo looks the way she does harken back to the early days of Neanderthal research. Following the 1856 discovery of a fossil specimen in Neander Valley, natural historians fell into two schools of thought as they tried to make sense of the Neanderthal skull—they had to decide whether the fossil represented a variant of humans or some other species entirely. Some said yes, the variation and cranial morphology were clearly different from that of *Homo sapiens*. Others argued that the differences in cranial capacity were easily explained by pathological variation and ascribed the differences as belonging to a malformed or diseased Cossack soldier. One explanation offered a new species; the other, a schema that explained morphology through pathology.

The hobbit seemed to replay this ideological divide. In other words, the same confusion, the same controversy, the same explanations that had already been introduced to explain an earlier paleoanthropological discovery offer a subtle reminder that types of discoveries and types of explanations maintain strong ties to their historical roots, and the explanations around the *Homo floresiensis* bones almost exactly mirrored the early explanations for differences in Neanderthals. The bones either belonged to a new species or represented the remains of a diseased and deformed individual. Most of the scientific community remains confident that Flo does indeed represent her own species.

The first question that confronted paleoresearchers was about Flo's size: Why was she so small? The authors of the original publications of *Homo floresiensis* argued that Flo represented an entirely new species of hominin, one that could have had a *Homo erectus*–like common ancestor, and that the extremely diminutive skeletal structure that Flo exhibited was simply the result of "island dwarfing" reflected in other lineages of island mammals (say, elephants and hippos), which evidence the reduction of a species' size over generations. Paleoanthropologist Dr. Dean Falk argued in 2005 that Flo's brain did not show evidence of microcephaly (an individual pathology), but that its shape meant it was not simply a "dwarfed descendent of *Homo erectus*"— rather, it indicated that it shared a yet-to-be-discovered ancestor. In other words, this undiscovered ancestor would serve as an evolutionary go-between between Flo and older species of *Homo*. The findings were challenged in a 2006 commentary in *Science* when the Field Museum's Dr. Robert Martin suggested that a reduction in body size from a *Homo erectus*–like ancestor would not result in "island dwarfing." Instead, Martin and others argued that for Flo to look the way she does, she would have had to be affected by several severe pathologies like microcephaly. (Microcephaly describes a suite of neurological abnormalities that result in small brains.) Most scientific consensus now, however, agrees that the process of island dwarfism best explains Flo's diminutive stature both in body structure and in brain size.

Studies of Flo's wrists and ankles show that her skeleton oscillates between primitive and derived characteristics; in some ways she is very similar to *Homo sapiens*, while in other ways she is very different. Although Flo's brain is small, evidence from studies of her wrist morphology demonstrate that she and others in her species were able to make and use stone tools. (Flo's brain size is measured at 400–426 cubic centimeters in volume, depending on the study cited, compared with approximately 1,300–1,350 cubic centimeters for modern *Homo sapiens*.) The stone tools—the artifacts that had sparked the early research for the *Homo floresiensis* team—indicated that Flo's species could hunt small elephants and large rodents. The presence of bones from hunted animals in ashes suggested that *Homo floresiensis* had a keen mastery of fire.[10]

Taken together, these characteristics provide an interesting evolutionary narrative. Geochronologist Dr. Bert Roberts of the University of Wollongong wryly noted that the increased complexity of the human evolution story immediately impacted everyday inquiries in paleoanthropology. "We had such a nice simple story, where we had modern humans and Neanderthals, and we bumped them off, that was the end of Neanderthals. We ventured across southeast Asia and it was basically empty because *Homo erectus* had died out there already, and we sort of just wandered into Australia and there we go. It was a clean and almost crisp little story. It made nice sense. Everyone was happy with that. And then suddenly the hobbit pops its head up."[11]

In 2006, an editorial in *Nature* bemoaned that excavations at Liang Bua had been suspended after tempers flared over the Flores remains. "The latter-day hobbit story was spiced up considerably by the characters of the scientists who discovered it—some of whom publicly and not always politely disagreed with one another about the discovery's significance. Not to mention the lively rebuttals from many academic challengers . . . who contend that to brand the Flores creature as a distinctive species is to create something as fictional as anything invented by Tolkien. Strong words have been exchanged. Skullduggery has been alleged. Accusations fly. This is ideal fodder for journalists."[12]

In order to better understand a fossil's evolutionary history, researchers need to be able to examine fossils—consequently, paleoanthropology is a science that depends on access. Access to collections, access to measurements, access to methods, and, of course, access to hominin fossils themselves. The question of what "access" means and how that translates into "good science"—a trade-off of collaboration and control—has been asked and answered many times throughout paleoanthropology's history.

In 2004, Teuku Jacob, chief paleontologist at the University of Gadjah Mada, a major power in Indonesian paleoanthropology and a proponent of the microcephalic argument, physically hauled off the specimens and had Liang Bua closed to researchers. Nominally, the bones were moved from Jakarta, the capital, to Yogyakarta, where Jacob's laboratory was located, in order to make casts. When the fossils were returned to Jakarta—months later than agreed upon—damage to the pelvis and jaw was clearly visible. An incisor from the jaw was missing, for example, and the jaw was broken in several places, thus making the reconstruction of the jaw very different than it had been before. Morwood and Brown claimed that the damage had occurred during the casting process; Teuku Jacob alleged that the damage had occurred before he and his laboratory took possession of the bones. It was only with Jacob's death that excavations at Liang Bua have been renewed. But for the Flores specimen, the question of access—and control to that access and interpretations of the fossil based on access to it—has become simply one more of many controversial aspects of the fossil's story.

Questions of fossil curation and who could—or ought to—examine the bones had huge implications for the everyday work of paleoanthropological science. How should discoveries like this be curated? Who ought to have access to fossils? And what kind of authority ought an expert's opinion convey? A good part of the controversy about the specimen can be attributed to the treatment of the bones and the actual access—or not—that surrounded the specimens.

Before Flores, as Brown put it, the "broad pattern of human pa-laeontology" had started "to look predictable." After Flores, as Mor-wood admitted, "challenges" arose to "existing notions of what it is to be human the most."[13] Additionally, there was discord between sev-eral research teams and several researchers about what the remains constituted: Were the remains a new species? Or did the remains represent a modern human from the late Pleistocene who simply had problematic pathology? "It is the most extreme hominin ever discov-ered," anthropologists Marta Mirazon Lahr and Robert Foley argued. "An archaic hominin at that date changes our understanding of late human evolutionary geography, biology and culture. Likewise, a pygmy and small-brained member of the genus *Homo* questions our understanding of morphological variability and allometry—the rela-tion between the size of an organism and the size of any of its parts."[14]

Many researchers have found themselves practically miring under the auspices of their own research agendas. In particular, Drs. Maciej Henneberg, Robert B. Eckhardt, and John Schofield published their book *The Hobbit Trap: How New Species Are Invented* specifically to argue that Flo isn't a new species, but simply a pathological variation in a modern human. Most in fact have come to accept *Homo floresiensis* as a legitimate species and not some pathological aberration, but that doesn't mean the debate has gone away. At the ten-year anniversary of the species' discovery, retrospectives, think pieces, and a renewed curi-osity about Flo's significance only intensified the debate. Staunch oppo-nents of the "hobbit species"—Henneberg and Eckhardt, together with colleagues Drs. Sakdapong Chavanaves and Kenneth Hsü—argued that LB1 was an individual with Down syndrome, claiming that this diagno-sis would explain Flores's skeletal morphology. But their insistence on diagnosing Flo's pathology has yet to resonate with the larger scientific community. Immediately after their publication, insults were bandied about and umbrage was taken. In short, Twitter had a field day.[15]

———

To complement *Homo floresiensis*'s scientific name, researchers cast about for a nickname as a way to introduce the public and nonspecialists to their discovery. The discovery was officially published in 2004, the same year that *The Return of the King*, the third installment of *The Lord of the Rings*, won an Oscar for Best Picture, and the fossil rode swiftly on the heels the hobbit movie craze. Although efforts have been made to shift the female specimen's nickname and identify it as "Flo" or "The Little Lady of Flores," for better or worse, the "hobbit" nickname has stuck. And little wonder why.

Dr. Roberts described the process of nickname hunting: "We knew we had to come up with a name for publicity purposes. We couldn't call it *Homo floresiensis*, so Mike said, 'I like hobbit.' I said, 'Okay as long as it's not going to cause any problems with Tolkien's estate,' or whatever they're called. They can get pretty stroppy with people using their trademarked words. Mike referred to LB1 as hobbit, not 'the' hobbit, as if its name was Mary. For a while, Mike was trying to persuade Peter Brown to call it *Homo hobbitus*. I think he just thought Mike was a complete charlatan for even suggesting it." (In *A New Human: The Startling Discovery and Strange Story of the "Hobbits" of Flores, Indonesia*, Morwood does indeed refer to the specimen without any article, just as "Hobbit.") Peter Brown added, "Mike and I didn't agree about nicknames because I thought it trivialized it, and I thought it would result in every loon on the planet telephoning me as soon as it was published. And that was true—endless bizarre telephone calls from people who had seen some small hairy person in their backyard."[16]

But there's a cultural power that comes from Flo's life and tenure as a "hobbit." As corny as it feels, referring to Flo as a "hobbit" reminds us that science doesn't act in a cultural vacuum. Mapping Flo's story onto something as familiar as a character out of a blockbuster story gives audiences a familiar "guide" to the fossil species. Pairing the fossil so closely with a well-known literary character has allowed the fossil to slip easily into the public's consciousness. The fossil has also risen to be part of Indonesia's national identity, implicitly offering the country a long-reaching historical narrative. But it's the controversy surrounding *Homo floresiensis* that has done more to define the fossil as a celebrity than any nickname. Apart from personalities and competing institutions, the Flores fossil had the opportunity to act as a nation-building symbol, similar to how Lucy served Ethiopia decades prior. It was "very important for Indonesian society," affirmed Professor Raden Pandji Soejono, the lead archaeologist at the dig.[17] Flores did for Indonesia, a recent democracy, what Hadar did for Ethiopia. Flo is a cultural artifact—a symbol—of nationalism, playing a role similar to Lucy.

Acting as more than just a literary trope, however, Flo has managed to resonate deeply through her cultural ties in Indonesia through stories about the ebu gogo, argues anthropologist Dr. Gregory Forth. According to the indigenous Nage of Flores, the ebu gogo are human-like creatures that live deep in the Indonesian forests. Forth has pointed out that calling the fossils "hobbits" is thus more than a simple literary allusion. "In particular, it has been found appropriate—evidently in order to communicate effectively with a wider public—to portray *Homo floresiensis* as a 'hobbit' (a choice obviously influenced by the recent Hollywood film versions of Tolkien's novels)," Forth argued in 2005 when the fossil was first gaining its scientific and public footing. Forth also pointed out that *Homo floresiensis* was a curious mix of anthropology—culture met biology square in the middle of Flores where legends like ebu gogo met scientific categories like *Homo floresiensis*. Not only does the name—nickname, even—show us particular cultural tropes through literary allusions, but the evidence used to talk about "how to know" about the species reflects our cultural backgrounds and assumptions as well. How we talk about the species and how we name it conveys how we ground that scientific discovery within our own cultures.

"Curiouser still," Forth continued, "the designation was not a creation of the popular press, but of the scientific discoverers themselves. Bound up with this identification, which has inevitably resulted in a trivialization of the anthropological discovery, has been a transformation of Flores into an approximation of Conan Doyle's 'lost world,' once the abode of pygmy elephants and still the home of giant lizards and giant rats (references respectively to *Varanus komodoensis*, or the 'Komodo dragon,' and the endemic Flores giant rat, *Papagomys armandvillei*), and perhaps even of dwarf hominids."[18] It is particularly easy to talk about evolution in literary tropes because these tropes offer narrative explanations to evolutionary phenomena. (One of the reasons that science fiction embraced the Neanderthal.) And as literature writes big narratives and big conflicts, the hobbit story and its controversies grow larger than life, visualizing phenomena outside of ourselves.

Homo floresiensis skull cast, juxtaposed with cave. *(Science Source)*

The hobbit species hasn't amassed the same kind of material culture or ephemera that we've come to ascribe to and expect from other famous hominin fossils. There are (as far as I know) no T-shirts, magnets, posters, or kitschy trinkets that surround Flo, the way there are for Lucy, Sediba, and even Neanderthals like the Old Man. But reconstructions of Flo, especially those by John Gurche, do live rather vibrant museum lives. Originally commissioned as a reconstruction for a National Geographic TV show, Gurche's Flores hominin now lives in the Hall of Human Origins at the Smithsonian. As part of the Hall's exhibit, next to a reconstruction of Lucy and a Neanderthal, Gurche created a bronze sculpture, catching Flo in a moment of panic, her dreadlock-esque hair flying around her face, her nostrils flared, and her arms outstretched—a moment of evolutionary pietà, fueled by the viewer's awareness that Flo doesn't, or can't, survive whatever she is shielding herself from. Her terror has been described as "biblical" and "classical."

Gurche's second reconstruction is a colored latex sculpture where a few strands of gray hair wisp around Flo's face and her soulful eyes seem to track the museumgoers throughout the exhibit. Both pieces emphasize Flo's evolutionary vulnerability. Some of Gurche's early sketches show Flo holding her head in her hands with her eyes shut, or with her hands thrown up over her head as if to stave off the inevitability of her extinction. For most reconstructions and dioramas, viewers are encouraged to believe that they are "seeing" a moment frozen in time. But like the dioramas of South Africa's Ditsong Museum, the slice-of-time sentiment is a convenient and necessary fiction to fold viewers into the scene and have them believe the narrative they see unfolding before them.

Gurche was curious what the effect of such a role reversal would look like, particularly with a hominin like the hobbit. "There are ways of playing with realism, of taking it a little bit further," Gurche muses. "What expression would be fitting for Flo if she were able to see us? What would she have thought of our kind?" In creating the expression and tone for Flo's headshot, Gurche drew on one of *National Geographic*'s most iconic covers—the photo of a green-eyed Afghani girl. In the photo, her expression is haunting, resigned, frightened, even distrustful. It's a lifetime of hardship focused through her eyes. "This is exactly the kind of expression I was thinking of for Flo," Gurche recalls. "I thought her expression should be an uneasy one at the very least, maybe stopping just short of profoundly disturbed."[19] The story that Flo conveys through these reconstructions is an empathetic narrative of harsh evolutionary inevitability.

––––––––

The issues and arguments surrounding the fossil have been played and replayed, to the point that debates about the specimen's interpretation as well as its handling have simply become de rigueur. These controversies only fan the fossil's fame. ("Scientists at War over the Flores Hobbit Man Fossils," screamed an August 16, 2014, headline from the *Guardian*, fairly typical of *floresiensis*-based articles.) There is always a sense of provoking and wanting to stir up disagreements about the fossil. Dean Falk recalled the day that *Nature* lifted its media embargo on the *Homo floresiensis* papers and authors were no longer barred from talking about the fossil discovery. Falk received a call from *National Geographic*'s David Hamlin. "While we continued to talk, I called up a news site on my computer and watched in amazement as Hobbit stories popped up one after another."[20] News about the discovery became weighted with an immediacy that other celebrity fossils had not been fraught with, thanks to the easy availability of information due to easy digital access. (No other fossil had been able to digitally go viral before *Homo floresiensis*.) Articles begat other articles, journalists roved the Internet looking for quotes, and blogs went wild. It was a far cry from the publication of the Taung Child or even the press release that accompanied Lucy's discovery.

Although originally a bit skeptical of the Flores discovery and its implications, Peter Brown later mused, "Now I'm more open to the idea that very small-bodied and small-brained bipeds moved out of Africa at a much earlier date, maybe 3 million years ago, or earlier. I'm more open to the idea that there were lots of failures in the evolution of bipeds. Some were successful, some weren't. It's a very branchy tree, and it just so happens we've survived."[21] Bert Roberts summed things up this way: "To me, the ultimate value of the hobbit is not what it is, in and of itself, because it's just a dead end. It probably didn't lead to anything that's alive today. But it opened up the door for people to think more broadly about everything. I think the hobbit changed the way people thought."[22]

"*Homo floresiensis* challenges us because she is so unexpected, because she does not fit with many preconceptions about how humans evolved and behaved, and what they should look like. Taken in context . . . she is, however, exactly what we might expect," Morwood argued in 2007, offering his take on why the fossil is such an unexpectedly expected conundrum. "Some find this possibility not to their liking and have challenged it, which in turn has led to a sometimes bizarre series of twists and turns in Hobbit's post-excavation history."[23]

Although controversy has so far defined the life of Flo the hobbit-like hominin, it is perhaps unsurprising, given how relatively recent the discovery is. If one were to look at only the first ten years of the Taung Child's life postexcavation, one would find a fossil defined by controversy. It wasn't until several decades later—and, especially, not until the debunking of Piltdown—that the Taung Child became less controversial and more mainstream. Many fossils in paleoanthropology's history have become well known rather rapidly—usually thanks to something provocative or heavily debated within the field—and then the fossils flame out, celebrity-wise, within a couple of decades.

Flo is a celebrity fossil for two main reasons. First, she is surrounded by controversy, and that controversy has spurred on scientific and popular recognition. The second reason for her celebrity has the possibility to keep her in the public's eye and to make sure that she resonates on more than just a level of scientific controversy. Because her biology and history—her small stature and moment of discovery—aligned so perfectly with the early twenty-first-century *Lord of the Rings* spectacle, she has a cultural resonance that gives her more footing than other fossils. *Homo floresiensis* even worked its way into popular television—references to the discovery pop up in season five of *Gilmore Girls*; in *Bones*, Dr. Brennan and her graduate student Daisy Wick traveling to Flores to look for more "hobbit" specimens.

With a different evolutionary story—if she weren't quite so small, if she had a larger brain, if she hadn't died so relatively recently—and a different cultural context—if the world hadn't quite taken leave of its senses over the Tolkien novels coming to film, if the scientific community hadn't become so publicly embroiled in its own discord—Flo's story would be completely different. Other fossils have been controversial and famous for it, but controversy alone doesn't make or sustain a celebrity fossil discovery in the decades after its discovery. She doesn't easily fit into either traditional phylogenetic narratives or cultural ones. In essence, she flipped evolutionary history on its—small—head, by inverting the paleocelebrity story. For fossils like the Taung Child and the Old Man, popular culture spent decades catching up with its then contemporaneous scientific discoveries. For Flo, popular culture had a neatly carved-out cliché—the hobbit—that she so neatly fit into. What this inversion means for Future Flo will be determined in subsequent decades. Flo might end up as a museum and national icon like Lucy, or she might end up buried in the cultural equivalent of a cave on a remote island of history, if *Lord of the Rings* can't sustain her in the decades to come.

If, at the end of the day, controversy is all that defines her, the odds are that she won't be famous in forty years—she'll simply be a curious footnote in paleoanthropology's history. Give the fossil another fifty years, and its cultural history will be richer, deeper, and simply different than it is now.

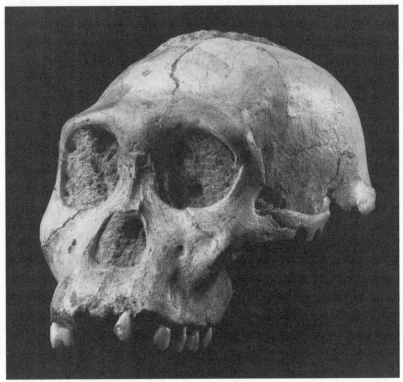

Portrait of *Australopithecus sediba*, or Karabo. *(Photo by Brett Eloff. Courtesy of Lee Berger, University of the Witwatersrand; CC GFDL)*

CHAPTER SEVEN

SEDIBA: TBD (TO BE DETERMINED)

D ad, I found a fossil!"

On August 15, 2008, nine-year-old Matthew Berger tagged along with his father, paleoanthropologist Dr. Lee Berger, on a field project in Malapa Nature Reserve in northern South Africa. The project was part of efforts to explore and map out known fossil sites and caves in the reserve, about forty kilometers north of Johannesburg. While puttering around the reserve with his dog, Tau, Matthew discovered what he knew to be some kind of fossil sticking out of a dark brown chunk of breccia rock. At first glance, the senior Berger thought that the fossil was simply a piece of a very, very old antelope—a common fossil in the area.

He picked up the block of rock containing the fossil and looked more closely, and realized that what he was looking at was a clavicle— a collarbone—of a hominin. He flipped the block over and saw a lower jaw encased in the same piece of breccia. "I couldn't believe it," Dr. Berger giddily recalled in a *New York Times* interview. "I took the rock, and I turned it [and] sticking out of the back of the rock was a mandible with a tooth, a canine, sticking out. And I almost died. What are the odds?"[1]

––––––

In April 2010, the fossils Matthew and his dad's team discovered in excavations from Malapa were published in *Science* as a new fossil hominin species called *Australopithecus sediba*. Although the paleoanthropological community was basically in agreement that the fossils were truly spectacular specimens, the scientific name proved to be a somewhat controversial taxonomic assignment because the fossils showed primitive apelike traits as well as derived, or *Homo*-like, characteristics. (Many researchers argued that the anatomy of Sediba would be better ascribed to the genus *Homo*, not to *Australopithecus*.) The publication of the fossils was accompanied by numerous opinion pieces arguing about the best taxonomic status for the fossil—from *Science* to *Nature* to *National Geographic* to the *New York Times*.

Regardless of its taxonomy, to date, the Malapa site was undeniably a significant fossil locale, having yielded over 220 bone fragments that, when put together, can boast a total of six skeletons: a juvenile male, an adult female, an adult male, and three infants that all lived around 1.9 million to 2 million years ago. When the fossil species was described in 2010, it was—and still is—tremendously exciting not only because Sediba lived during a time when both australopith species and early *Homo* roamed the greater African landscapes together, but also because the fossils were from multiple individuals with incredible archaeological provenience. These fossils represented an interesting time in our evolutionary history and constituted a sample of the species that was greater than just one individual—which, in turn, helps paleoanthropologists understand variation within fossil species.

Over the twentieth century, little did more to shape paleoanthropology's emerging identity as its own scientific discipline than the fossil hominin discoveries from Europe, Africa, and Asia. Every new discovery inherently carried a certain prestige because the fossil discoveries offered the basis for creating hypotheses and explanations about what could be observed in the fossil record—new fossils could make or break definitions of species, and every new discovery had the potential to rewrite the family tree. New fossils were imbued with social prestige in their original contexts—either accepted as ancestrally significant, like Peking Man, or dismissed, like the Taung Child.

As more and more fossil discoveries have entered the scientific record over the course of the last century, fossil collections are simply not as sparse as they were in earlier decades. (There are, for example, over four hundred Neanderthal individuals represented in the fossil record so far, compared with the very few specimens of the nineteenth century.) So, where does this leave twenty-first-century fossil discoveries? What would a famous fossil look like today? Flo and *Homo floresiensis* gave us one type of modern celebrity—contentious little hobbit that she is. The discovery of Sediba raised other questions: What historical patterns could or would other fossil discoveries follow? What historical patterns *would* they follow? What cultural expectations—and what scientific questions—would twenty-first-century fossils now need be required to answer to?

"The dolomitic cave deposits of South Africa have yielded arguably the richest record of both hominin and mammalian evolution in Africa. Fossils were first recognized in these deposits in the early 20th century, but it was the discovery of the Taung child skull from the Buxton Limeworks in 1924 that led to the recognition of the importance of these cave sites," Berger explained in a guide to the fossils and history of the Malapa region.[2] Part of the reason that the Malapa specimens could catapult so quickly into the paleo limelight was due to the incredible paleoanthropological history associated with the Malapa—Sediba's success is contingent, in no small part, upon the fossils' South African legacy.

But Sediba's renown is also a product of the fossil being in the right place at the right time and with a person to champion it, all the while pushing for a change in the paradigm of how paleoanthropology collects data and generates hypotheses. If the historical parallels are any indication, the life and afterlife of a fossil are made and remade by its contexts; its lasting celebrity is created over decades. While Sediba's initial life history certainly sets it up to be The Next Big Thing, it's not a foregone conclusion that a century from now it will still carry the same distinction it has today.

———

The year 2010 was a great one for studies in human evolution: two major new hominin discoveries entered the scientific record. These new fossils were both members of the genus *Australopithecus*, around two million years old. Although both were significant in terms of broadening our understanding of hominin evolutionary history, the two fossils have had very different lives after their discoveries. As similar as the fossils appear at first glance, they actually couldn't be more different. One was from South Africa, the other from Ethiopia; one was published in the *Proceedings of the National Academy of Sciences*, the other in *Science*. One was an *Australopithecus afarensis*—like Lucy—and the other was the new species, *Australopithecus sediba*. One was a discovery of a partially complete single skeleton, and the other featured multiple individuals. One was discovered during a routine field season by a veteran member of an international research team, and the other by a nine-year-old boy and his dog. The one, Kadanuumuu, has languished in scientific journals as a rather unknown hominin specimen to the public, while the other, Sediba, has gone on to wide international acclaim. Both are unquestionably significant to the field of paleoanthropology, but the cultural lives of these fossils are as separate as their evolutionary trajectories.

But why? What makes one fossil famous and not another? Why did one capture public and scientific attention and the other not?

The short answer is easy: context. Not only did the fossils come from different geological contexts—Kadanuumuu from East Africa's Rift Valley and Sediba from northern South Africa's limestone caves—but, more important for our purposes, the two sets of fossils inherited their own context for the history of their science, their own research traditions, and their own regional histories of how fossil discoveries are written into the story of human evolution. These differences came to speak volumes in how the fossils are studied and how they are immortalized.

The longer answer is, of course, more complicated. As with many famous fossils—Piltdown, Peking Man, Lucy—scientific significance is certainly *one* reason for fame and celebrity, but it is not *the only* reason. Paleoanthropology is a science punctuated by discovery and built by the fossils that it finds. Fossils capture scientific and public imaginations as new discoveries fill media headlines and Twitter feeds. These two fossil discoveries play out as brilliant cultural foils for each other—they are such recent discoveries that their initial conditions are easily compared and contrasted. From the same starting point—the same year of publication—they tell two different stories, giving audiences what-if scenarios for how twenty-first-century fossils enter scientific and social circles.

On his most basic level, Kadanuumuu—or "Big Man," as he is affectionately called by the research team—is a partially complete skeleton assigned to the species *Australopithecus afarensis* and designated by field catalog number KSD-VP-1/1, dating to 3.5 million years ago. The first element of the partial fossil skeleton discovered was the proximal part of an ulna—the part of the forearm that makes the elbow joint—on February 10, 2005, by Ato Alemayehu Asfaw, an established member of the international team of paleoanthropologists. In addition to the ulna, the rest of the Kadanuumuu skeletal materials were published five years after its initial discovery in the *Proceedings of the National Academy of Sciences*, a respected top-tier scientific publication. The article's authors represented an international team, with members from the Cleveland Museum of Natural History, Kent State University, Case Western Reserve University, Addis Ababa University, and Berkeley Geochronology Center.

In their publication, the authors noted that the fossil was "extraordinary" because the discovery expanded the knowledge base of *Australopithecus afarensis*. Most significant was the information it provided on how the species walked. In the decades since her discovery, there have been extensive debates about the exact nature of Lucy's bipedality; sure, she could walk upright on two legs, but just how much of her time was she bipedal, and how efficiently did she walk? How much was she like us? With the skeletal elements recovered, Kadanuumuu was able to refine and answer questions that focused on how the *afarensis* species would have moved.

Kadanuumuu also sported a complete scapula—part of the shoulder—which meant that scientists could examine if and how *Australopithecus afarensis* moved in trees and how it could have moved its shoulders. In an interview with *Nature*, scientists offered their take on why this new information was significant. "This new skeleton shows a fully running and walking biped, with most of the adaptations we have," said team member Dr. Owen Lovejoy, a paleoanthropologist at Kent State University. "What we see in the new skeleton's pelvis is what we see in modern humans," added the article's lead author, Dr. Yohannes Haile-Selassie of the Cleveland Museum of Natural History. Science writer Rex Dalton described the discovery this way: "A hominid species made famous by the 'Lucy' fossil from Ethiopia could walk down a runway just like a fashion model today, a newly reported partial skeleton shows."[3]

Even in Kadanuumuu's original publication, he lived in Lucy's shadow. The team referred directly to Lucy in the second sentence of the article's abstract, and the article's photograph of Kadanuumuu is the classic allusion of Lucy's iconic portrait—bones laid out in anatomical

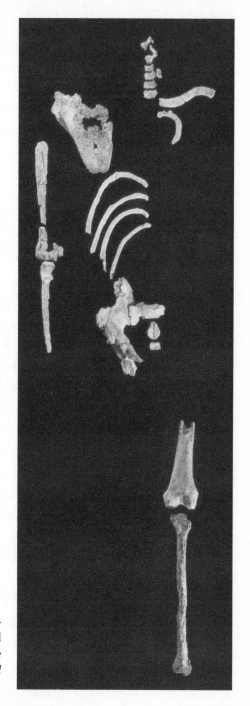

Portrait of Kadanuumuu, *Australopithecus afarensis*, published in 2010. *(Yohannes Haile-Selassie and Cleveland Museum of Natural History. Used with permission)*

position against a black background. Where Kadanuumuu's "official portrait" suggests Lucy, it also subtly highlights the differences between the skeletons. Lucy has some skull fragments and a jaw in addition to her long bones—so it's easy to mentally fill in the parts that are missing, or at least grasp how she could have been a living entity. Headless Horseman–like, Kadanuumuu has no crania and only one leg. In other words, with certain skeletal elements, like a skull, it's easier to anthropomorphize certain fossil specimens; the easier to anthropomorphize, the easier to create a character that people will identify with.[4]

Even at his press release, the official description of Kadanuumuu relied heavily on the audience's familiarity with Lucy. "The new skeleton comes from the Rift Valley in the central Afar of Ethiopia, about 330 kilometres northeast of Addis Ababa," Dalton noted. "Found in 2005 . . . a long day's walk north of Hadar where Lucy was discovered . . . the skeleton is estimated to be nearly 2 metres tall. Lucy was just over 1 metre tall."[5] A short 2010 piece in *National Geographic* titled "'Lucy' Kin Pushes Back Evolution of Upright Walking" supplies similar information about Kadanuumuu's morphology and skeleton, but once again defines the fossil against Lucy. A 2015 study that examined the question of sexual dimorphism—the physical differences between males and females of the same *Australopithecus afarensis* species—again juxtaposed Kadanuumuu with Lucy. This would be expected given how significant Lucy is—but what really undercut Kadanuumuu was the study's emphasis on Lucy. Her name was first in the study's title, weighting the better-known fossil over the newer, less-studied one.[6] The Kadanuumuu fossils expanded the details of how we think about a fossil species—especially how we think about interspecies variation and the nuances of locomotion for *Australopithecus afarensis* hominins. But for all the emphasis on the fossil's bipedality, culturally speaking, Kadanuumuu has trouble standing on his own two feet.

Why? Because Kadanuumuu is a semicomplete *Australopithecus afarensis* whose origin story is fairly de rigueur for paleoanthropology. There isn't anything about the fossil, its discovery, its science, or its museum life that really jumps out and grabs audiences—nothing that revises the hominin phylogenetic tree or that inspires an entirely new canon of scientific inquiry. Kadanuumuu does not represent a new species or a new archetype. He doesn't represent a new set of questions to paleoanthropology and he doesn't highlight really new methodologies.

Kadanuumuu is a fossil—much like Turkana Boy or Mrs. Ples—that, perhaps, will jingle some note of recognition in audiences, but will quickly fade. (Mrs. Ples is the adult *Australopithecus africanus* discovered in Sterkfontein, near Taung, by Robert Broom in 1947 that substantiated the Taung Child. Turkana Boy is a *Homo ergaster* specimen found by Richard Leakey near Lake Turkana in Kenya. Both are important discoveries in the history of paleoanthropology, but simply haven't risen to the celebrity strata of other fossils.) Kadanuumuu, like Mrs. Ples, props up other fossils. On the off chance that Lucy's luminous essence might diminish, Kadanuumuu is right there, ready to lend the little cachet that it has to offer; Kadanuumuu exists as a secondary character to Lucy—a member of the *Australopithecus afarensis* supporting cast, her understudy. It's as if Kadanuumuu is a bit player that you sort of recognize on TV, but it takes three clicks on Wikipedia to remind yourself why you recognize the actor. It's difficult to become a celebrity fossil living in Lucy's shadow. It's important to note that not all researchers necessarily *want* fossils to become famous; decent, respectable, significant science can and certainly does come from fossils that never quite crack into popular imagination by either chance or choice.

———

Sediba is a fossil of a different sort, and the story of his social celebrity is completely different from Kadanuumuu's. First off, one of the huge differences between Sediba and Kadanuumuu is in the fossils' names—both culturally and scientifically. *Australopithecus sediba* bucks the trend of famous fossils colloquially known through popular nicknames. While most famous discoveries rely heavily on the cultural staying power of a nickname—Lucy, the Taung Child, the hobbit—having a strong nickname is by no means a necessary step to celebrity. (Other fossils—Piltdown Man, Peking Man—are certainly simply informal shorthands for the fossil. Davidson Black even suggested calling the first hominin from Zhoukoudian "Nelly" to combat what he saw as explicit sexism by referring to particular hominin discoveries as "man.") This isn't for lack of trying for a cute name, though, as a nickname helps a fossil better connect to its public.

A press release issued by the University of the Witwatersrand in 2010, just after the formal publication of the discovery, raised the issue of a nickname and suggested that the fossil ought to have one. "The site continues to be explored and without a doubt there are more groundbreaking discoveries to come forth," the article reads. "In celebration of this find, the children of South Africa have been invited to develop a common name for the juvenile skeleton."[7] The skeleton, which is type specimen MH1, was eventually named "Karabo" ("The Answer") by Omphemetse Keepile, a seventeen-year-old student from Johannesburg. In *The Skull in the Rock: How a Scientist, a Boy, and Google Earth Opened a New Window on Human Origins*—a *National Geographic* children's book by Lee Berger and Marc Aronson—the Sediba skeleton is named Karabo, but that name is never used as a persona or shorthand for the fossil, indicative, perhaps, of how the nickname doesn't seem to have struck a chord. The fossils are known colloquially as Sediba instead.

Sediba is, of course, a shortened form of the fossils' taxonomic assignment *Australopithecus sediba*. The specimens are also referred to by their catalog numbers—MH1 and MH2—or, collectively, as the Malapa hominins. In keeping with paleoanthropology's emerging twenty-first-century tradition of tying fossil names to local languages, the word *sediba* comes from South Africa's Sotho language. *"Sediba, which means natural spring, fountain or wellspring in Sotho, one of the 11 official languages of South Africa, was deemed an appropriate name for a species that might be the point from which the genus Homo arises,"* remarked Berger. "I believe that this is a good candidate for being the transitional species between the southern African ape-man *Australopithecus africanus* (like the Taung Child and Mrs. Ples) and either *Homo habilis* or even a direct ancestor of *Homo erectus* (like Turkana Boy, Java man or Peking man)."[8] Tying its name to its region of discovery, Sediba triangulates the geography, taxonomy, and evolutionary narrative that are implied through its scientific name, *Australopithecus sediba*.

The fossils were undeniably something new for paleoanthropology, but their discovery was, perhaps, not completely unexpected given their geological provenience. The Malapa cave site is part of an area of northern South Africa called the Cradle of Humankind, designated a UNESCO World Heritage site in December 1999. The Cradle is part of a large geologic complex of limestone caves, approximately 7,000 hectares or about 180 square miles in size. For close to one hundred years, the Cradle has continually yielded new fossils and new species that pique paleoanthropology's curiosity and offer unique possibilities for teasing apart and putting together an evolutionary narrative.

"Hominins are represented in the South African cave sites by over 1,000 catalogued specimens from more than 11 different cave deposits. . . . At least four and possibly more species of early hominin are found in the South African cave sites," Lee Berger offers in his *Working and Guiding in the Cradle of Humankind*. Again, since the geological contexts between East and South Africa are so different, they offer different patterns for fossil discovery and offer evidence for different time periods and geographic places in humanity's evolutionary story. "While the hominin fossils from South Africa are not nearly as old as the oldest hominin sites in East Africa (East African fossil hominins [like Lucy] may date back over six million years while those in South Africa are probably all less than three million years in age), the South African examples are important because they are almost always more complete and are found in the presence of a much greater range of vertebrates. They are therefore able to tell us a lot about the period in which they lived."[9]

One thing that sets the Malapa specimens apart from other fossils was the quick turnaround time between discovery and publication. While Kadanuumuu's publication was a very respectable five years after its discovery, *Australopithecus sediba* was published just two years after the discovery of the fossils. The initial 2010 *Science* publication, titled *"Australopithecus sediba*: A New Species of Homo-like Australopith from South Africa,"* was simply a warm-up exercise for the marathon of articles that Berger and his team would publish in the next half decade. In 2011 alone the Sediba team published five in-depth articles about the fossil in a special issue of *Science*, each article tackling a different anatomical element (pelvis, ankle joint, etc.) and one on the process to assign a geological date to the fossil.

People were charmed by the species—even its detractors and nay-sayers. As a fossil species, Sediba represents an interesting suite of anatomical characteristics. It has long arms, short powerful hands, a very advanced pelvis, and long legs. This mix of anatomical characteristics made it capable of striding, possibly even running, like a human. It is also likely that Sediba could have climbed. "It is estimated that they were both about 1.27 metres, although the child would certainly have grown taller. The female probably weighed about 33 kilograms and the child about 27 kilograms at the time of his death," added Berger. "The brain size of the juvenile was between 420 and 450 cubic centimetres, which is small (when compared to the human brain of about 1,200 to 1,600 cubic centimetres) but the shape of the brain seems to be more advanced than that of australopithecines."[10]

"These fossils and many others are landmark discoveries in

paleoanthropology, finds that have filled crucial gaps in scientists' understanding of human origins. They are all vitally important. And yet the *A. sediba* fossils manage to stand out from even this elite crowd, because of the sheer volume and quality of information they contain," argued science writer Kate Wong in *Scientific American*. "The finds from Malapa tick pretty much all the boxes on a paleoanthropologist's wish list. Specimens that preserve multiple skeletal elements? Check. Remains of multiple, coeval individuals (important for understanding variation within a species)? Check. Fossils in near-pristine condition, thus eliminating uncertainties about how pieces fit together? Geological context that allows for precision dating of the fossils? Associated plant and animals remains? Check, check, check."[11]

Wong's informal checklist offers several key points to begin to understand why Sediba has been culturally fast-tracked along its way to famous fossil status. However, simply marking off anatomic features and a successful archaeological context cannot, by themselves, generate a famous fossil. A famous fossil is more than simply the sum of its skeletal elements and more than the significance of its context; a successful celebrity fossil manages to gain traction outside of scientific circles and maintain a cultural persona. Lucy was the first mostly complete skeleton to enter the paleoanthropological record, but it was how her skeleton was used, viewed, studied, and written about that helped push her into a cultural context and give her a public persona. Sediba benefits from being the right fossil in the right time with the right discovery story and a scientist to champion it. Certain elements of its story have historical allusions to the Taung Child's discovery, and it benefits from a team savvy enough to leverage that history.

In many ways, the open, public access that is associated with the Malapa assemblage makes Sediba a very accessible fossil—both inside the scientific community and outside it. It's easy to talk about the Malapa specimens because it's easy to access them through publications, images, scans, and casts. "Many reviews of palaeontological research end with the statement that it would be highly desirable to recover more fossils. In this case, however, the Malapa team has already done that," argues paleoanthropologist Fred Spoor. "The interpretation of their findings may be a matter of debate, but they have undoubtedly added a spectacular and thought-provoking sample to the hominin fossil record. This achievement represents a major contribution to the study of human evolution in all its complexity."[12]

The question of fossil access is raised over and over in paleoanthropology. "The fossils are owned by the people of South Africa, and curated by the University of the Witwatersrand, Johannesburg," the original Sediba press release read. "They will be on public display at Maropeng in the Cradle of Humankind until the 18th of April 2010, will move to Cape Town for the launch of Palaeo-Sciences Week from the 19th of April and will again be on public display at the Wits Origins Centre during May, on dates to be announced shortly."[13] Not only were the fossils on display immediately following their publication, but casts of the fossils have been working their way through museum, popular, and scientific circles.

Berger's commitment to transparency and access goes beyond simply displaying the fossils or casts. When the Malapa fossils were being excavated from their breccia, Berger was quick to point out that he wanted to make the entire excavation available online and have nonexperts be able to interact with the scientists. In a 2012 interview with *National Geographic*, Berger's enthusiasm for the social life of the fossils was practically contagious: "The world is going to be able to watch and interact live as we expose this discovery. There's also the possibility that we have two bodies that are intertwined [in the rock]. Part of the fun of this project is that as soon as we find out, the world will find out with us."[14]

———

Ever since the fossil was published, it's been very approachable, thanks to the outreach efforts of Berger and his team. Images of

Sediba have flooded the Internet, and the fossil shows up everywhere, from scientific publications to museum exhibits to Wikipedia pages—the photographs, formal headshots, and snapped candid field shots help to tell a very visual story of Sediba, particularly the photo of the Malapa discovery itself, which is a candid shot of young Matthew Berger showing off the fossil while it's still in its breccia matrix. The photo has popped up everywhere from the Malapa site's Wikipedia page to articles in *Nature* to museum exhibits in Cape Town's Iziko Museum.

Matthew Berger and the Sediba fossil, still in matrix, at the Malapa Nature Reserve. *(Lee Berger; CC-BY-SA-3.0)*

Sediba's images, through photographs, casts, or reconstructions, are incorporated into a variety of public settings. Since the Malapa specimens are inexorably tied to the Cradle of Humankind, they feature prominently in the Cradle's visitors' centers.

As a UNESCO World Heritage site, the Cradle is heavily marketed in South Africa as a paleotourist destination. Its main visitor center is Maropeng, opened December 7, 2005, by then president Thabo Mbeki. Anthropologically, Maropeng offers visitors an opportunity to explore the region's fossils and human evolution as a whole. Architecturally, the building is covered in grass, rising like a gigantic gnome house out of the stark South African landscape. For the paleo-adventurous, the Maropeng center offers a boat trip through the "Tunnel of Time," where visitors comfortably float from the Cretaceous to the Pleistocene, passing through landscapes full of recorded pterodactyl screams and eventually tooling around Pleistocene volcanoes and ice floes; the Disney-esque boat ride ends at the Hall of Human Origins, which showcases hominins from around the world. All of the "rock star" fossils, from Lucy to Taung to Neanderthals to the Malapa fossils, are featured. At Maropeng and other Cradle museums, Sediba manages to move from a strictly scientific object to tourist ephemera—small 3-D-printed Sediba skulls are sold in gift stores as necklaces and key rings. It's easy to know about Sediba because Sediba is right there, ready to be known through casts, photos, tourist trinkets, and museum exhibits.

Even in more formal scientific settings, the images of Sediba have dominated over those of other fossils. Take the cover of *Science*. With its full-page picture and specialized typography, the journal's cover conveys intellectual gravitas and scientific legitimacy, and has for decades. Ever since *Science* introduced a picture as part of its cover in 1959, the publication has featured a plethora of images, from thin-section slides and meteorological phenomena to pollen spores and technical instruments. Sediba has graced the cover of *Science* three times since 2010, a feat unmatched by any scientific discovery in such a short amount of time—and unmatched, in fact, by any other fossil in the history of the journal's publication.

Only nine covers in over fifty years have featured hominin fossils. The first cover was fairly recent: a June 1998 cover showcased a color illustration of the two- and three-dimensional computer imaging of the endocranial capacity of Stw 505, an adult *Australopithecus africanus*. An August 1999 cover showed the forelimb bones and jaw of a partial skeleton (*Equatorius*), a very old specimen from a site at Kipsaraman, Kenya. The March 2, 2001, cover gave faces to the 1.7-million-year-old male and female hominins from Dmanisi, Georgia. The rather recently discovered fossil *Ardipithecus ramidus*, nicknamed Ardi, graced the cover twice in rather quick succession, in October and December 2009.

Unprecedented in *Science's* history, *Australopithecus sediba* was on the cover, on April 9, 2010, September 9, 2011, and April 12, 2013, each cover showing Sediba in different poses: one with the skull, one with Sediba's hand, and the final one a fully reconstructed skeleton with the left hand slightly extended, almost inviting the reader to join him. (The most recent paleo-related cover to date, October 18, 2013, was a photo of a 1.77-million-year-old complete adult skull from Dmanisi, designated as early *Homo*.) Each of these covers conveys the importance of the fossil discovery. Framed enlargements of Sediba's *Science* covers hang in the hall of the Evolutionary Studies Institute at the University of the Witwatersrand—like an agency proudly showcasing headshots of its successful models.

———

I had the opportunity to meet the Sediba fossils in person one summer at the University of the Witwatersrand. From the university's archives building, I hiked across campus to the Evolutionary Studies Institute located in the Palaeosciences Centre. Dr. Lee Berger cheerfully spent a

morning chatting about the Malapa project, the history of paleoanthropology research in the Cradle of Humankind, and the nature of celebrity fossils. The laboratory space itself is arranged to study all sorts of
fossils—not just Sediba. Long benches offer space for researchers to
examine fossils, both casts and the real ones; screen savers jogged across
monitors attached to computers crunching data; students and postdocs
chatted about their various research projects. That June morning, the
hum of conversation and activity filled the sunny lab room. It was
impossible to see the Malapa fossils and not conclude that they were
enjoying their well-won status in the paleoanthropological world.

A giant vault for fossils stood at one end of the lab. As Berger
deftly entered the combination, he talked about what he saw as the
amazing, still unexplored potential for South Africa to contribute to
big questions in paleoanthropology. (His insistence on this point was
justified. In October 2013, he and a team of researchers began excavating the Rising Star Cave, which produced over 1,200 hominin
bone fragments; subsequent excavations in April 2014 yielded 1,724
hominin fragments. The initial publication, in September 2015, described the hominins as a new species, *Homo naledi*.)[15] During my
visit to the Witwatersrand lab, Berger pulled out the cases that held
the Malapa fossils and set them on one of the laboratory tables. A
colleague of his, Dr. Steven Churchill, wandered over to join us.
Berger opened the crates and we peered down at the famous Sediba
specimens. Big bones, smaller bones, even some miniscule bone fragments. Each fossil was carefully nestled into its own foam insert, and
every insert was labeled with the specimen's catalog number.

A large scanner stood off in the corner, and Berger pointed out
how the team had used the machine to scan parts of the roughly
excavated fossil matrix from Malapa. Since large chunks of calcite
breccia rock, infused with fossils, had been excavated from Malapa,
the finer grain extraction of the fossils from the rocks happens at the
lab. With the help of the scanner, scientists are able to get an idea

about what is inside the breccia chunk before cutting into it, helping to better preserve the fossils.

Berger talked through each of the skeletal elements, highlighting different anatomical features and comparing them to other hominins, with Churchill occasionally interjecting or offering an observation. (Jokingly, Berger asked whether I, as a historian, was able to predict whether fossils could or would become famous.) From his discussion, it was beyond clear that he was completely committed to making the fossils available for study, whether through the originals or through casts. The passion for the new fossils—especially since they represented the discovery of a new species—was palpable. But even more profound was the sense that the project was *different* because the science going on around them was somehow different. Or at least, the science around the fossils was being done differently.

For a good chunk of its history, paleoanthropology has been a field dominated by precious few fossils and an implied hierarchy of knowledge based on access to those few fossils. Controlling who can look at what fossils and when has been a means of controlling what scientific and social narratives dominate the field. On a very broad scale, knowledge about human evolution is created by studies about the fossils—measurements, comparisons, statistical analyses; ipso facto, whoever controls fossils controls the production of the field's knowledge. This can mean either gatekeeping (keeping out the cranks) or guarding (preventing dissenting voices). Access cuts both ways.

Berger and his team, tired and even disgusted by the problem of access to fossils, have vowed to not let that happen with the Sediba specimens. "The way Berger and his collaborators are studying the finds and disseminating what they learn represents a real departure from the cloak-and-dagger manner in which paleoanthropological investigations often proceed," Kate Wong argued. "Berger has assembled a huge team of specialists to work on the remains and has made the project open access, with a policy of granting permission to any paleoanthropologist who asks to see the original fossils. He has also sent out scores of replicas to institutions around the world, and routinely brings casts of the bones—even ones that his team has yet to formally describe—to professional meetings to share with other researchers. This can only improve the quality of the science that comes out of the project and may well inspire other teams to be more forthcoming with their own data."[16]

There is palpable excitement over this "change" and what it means for the field. But this eagerness about the Sediba fossils raises the question of what change in a scientific field like paleoanthropology looks like, how to make sense of it, and what kind of results are reasonable to expect from changes to the way scientific knowledge is produced. Because that's what's really at stake here with fossils like Sediba—a challenge to the paradigm that knowledge must be produced from the top down, once access to fossils has been granted.

The question of change in science is certainly well explored and well studied in the history and philosophy of science. When looking at change in science on a grand scale, we see that the big idea-based changes happen as part of what historian of science Thomas Kuhn called scientific revolutions and paradigm shifts. Other philosophers and historians of science, particularly in the decades after Kuhn, argue that changes happen additively, slowly and over time—where new ideas and methods propagate almost evolutionarily, where research can be understood as a series of research problems, and where each problem is solved in order of its significance or importance to the field.

Here, in the first part of the twenty-first century, paleoanthropology has all the markers for huge changes within its discipline, and these changes are reflected in how newly discovered hominins are studied. Just as Taung illustrated historical shifts in paleoanthropological theory, fossils recovered from the Malapa site—and the subsequent Rising Star Expedition—can help us to consider new intellectual trends in methodology of the discipline, like publishing the fossils in a way that is accessible to a broader audience or even posting 3-D scans of the fossils themselves, inviting others, including nonexperts, to participate in the science-making process.

The Sediba fossils, in fact, represent a very clear change in how paleoanthropology opts to create knowledge but not necessarily engage with new research questions. They represent change in science thanks to tools for studying fossils, rather than big ideas. Where some, like Kuhn, assumed that new big ideas were the primary drivers of scientific change, others suggest that new tools and new methodologies are more apt drivers for change at the turn of the twenty-first century. This is definitely the kind of scientific change that Sediba represents: namely, paleoknowledge being generated from new methodologies (like new casting technologies or 3-D scanning and printing), and new approaches to fossil access, such as publishing in a timely manner and with easy, open access to fossils. These differences underscore the differences between Sediba and Kadanuumuu.

Excavation projects at Malapa and elsewhere in the Cradle seem to be modeling the process of creating knowledge after other "big sciences." In other sciences, such as biochemistry and physics, discoveries cannot be undertaken by a single person or research institute, as data sets are too large and experiments too complex. In paleoanthropology, recent changes include: increased access to fossils, accessibility of data, a transparency of methods, technology of casting and 3-D printing and dissemination, timely publications, and a public engagement. These new characteristics would seem to be a broad call for the discipline to reconsider how it "does science." The Rising Star Expedition, for example, was a direct outgrowth of the paleo fame and fortune imbued by the successes of Malapa and Sediba; we see people looking to create knowledge from a broader cohort of scientists, providing access to fossils, drawing on a variety of expertise, and offering transparency and accessibility to nonexperts (via blogs and Twitter) in how the processes of science work. The hope is to involve more people in the process of scientific knowledge making and to have that process be more transparent.

———

Sediba is a curious celebrity fossil, in part because its story is so recent and is thus still unfolding. If we build the Sediba fossil through comparison and contrast—particularly compared with Kadanuumuu—it's easy to see the impact that the initial conditions of the fossil's public and scientific life play a significant role. But one of the most interesting aspects of the fossil—perhaps part of the draw of it—is that almost every element of what makes a fossil famous can be found in Sediba, even with its relatively short postdiscovery life.

In an interesting twist, both lead authors—Yohannes Haile-Selassie and Lee Berger—have leveraged their fossil discoveries published in 2010 into additional discoveries. Haile-Selassie was the lead author on an article that described a completely new australopith species, *Australopithecus deyiremeda*, in May 2015, while Berger was the lead on the 2013 Rising Star Expedition, which began excavations at another cave not far from Malapa.[17] It's as if 2010 repeated itself in 2015: one discovery was famous—*Homo naledi*'s press releases and the tour of its fossils have saturated science media—and the other not so much. The contrasts sharpen the question of how much different types of discoveries are made famous by their stories and surrounding contexts. Are Berger and his team simply successfully using social media—live-Tweeting excavations and carefully maintaining comprehensive Wikipedia pages—where other discoveries are not? Or is one more famous through happy accidents of its context?

Within scientific circles, Sediba has been interpreted as a potential ancestor for the *Homo* genus; the species' morphology represents some apelike and some humanlike characteristics, which harkens back to many historical phylogenetic debates—Taung, Lucy, even the Old Man of La Chapelle. Sediba is also able to invoke these other narratives and other aspects of celebrity fossils. It's as if one were to take the best parts of the lives of Taung Child, the Old Man of La Chapelle, Peking Man, and Lucy and distill them into a single set of specimens. (The only famous fossil story that Sediba doesn't tie into is Piltdown and, it goes without saying, that's just as well.) With so many types of famous fossil discoveries in paleoanthropology's history, it becomes very easy to talk about new discovery in terms of how it compares with older ones. "What [Berger] has shown in South Africa is that when you work with the government to open access to things, that has huge benefits for the country," paleoanthropologist Dr. John Hawks suggests. "The amount of attention South Africa has gotten for Sediba is more than any other country got since Lucy. Such positive attention is hard to come by."[18]

A fossil like Sediba has all the "right" elements to continue on its trajectory to paleocelebrity. On one hand, Sediba's contribution to paleoanthropology is obvious—as a new fossil species in a complex evolutionary time in the hominin family tree, the fossils are well positioned to be studied for decades by countless researchers. On the other hand—on a more subtle level—Sediba is also well positioned to challenge the social worldview, the mechanical "doing" of science that serves as the basis of paleoanthropology as a scientific discipline. Kate Wong notes, "The strategy has paid off. Researchers have flocked to South Africa in droves to check out the remains, Berger's research team has grown to include more than 80 members, and within just a few years of getting the bones out of the ground the team has already published a raft of high-profile scientific papers, with more in the pipeline."[19]

Sediba's story continues to unfold, much like Flo's and other very recent discoveries. But Sediba's story also asks a lot of questions about the process of creating scientific knowledge that other fossils do not. It's apparent that Sediba has cultural cachet in spades. The next hundred years will determine just what kind of celebrity Sediba has to offer—it's still To Be Determined—but for the Malapa fossils, their celebrity feels imminent.

O FORTUNA!:

A BIT OF LUCK, A BIT OF SKILL

I n 1929, writer-philosopher Ayn Rand worked as an office grunt
for RKO Pictures' wardrobe department, three years after emigrat-
ing from Russia. It would be fourteen years before she published
The Fountainhead and twenty-eight years before *Atlas Shrugged*. In
1929, she was simply an aspiring novelist, writing short stories and
clocking in hours at the studio to pay the rent. Although Rand loathed
her time with RKO's wardrobe department, the hype and hyperbole
of Hollywood would provide the fodder for her short story "Her Second
Career," about a fictional movie star, Claire Nash.

To the outward observer, Claire Nash has all the trappings of a successful Hollywood movie career: a palace in Beverly Hills, two Rolls-
Royces, and the unending admiration of her thousands of fans for her
"sweet maidenhood" screen persona. ("For her, five gentlemen had committed suicide—one of them fatally—and she had had a breakfast cereal
named in her honor.")[1] Nash was considered the most brilliant actress in
Hollywood, a career that was the goal and envy of every aspiring starlet.

Winston Ayers, Ayn Rand's fictional playwright in the story, argues
with Nash over the notion that her success—her celebrity—is somehow

earned. "You see [screen actresses] are not one in a thousand, they are just one out of a thousand, chosen by *chance*. Thousands and thousands of girls struggle for a place in the movies. Some are as beautiful as you are, and some are more beautiful. All can act as you act. Have they a right to fame and stardom? Just as much or just as little as you have." Ayers then challenges Nash: "You have made your career. I do not ask how you made it. You are famous, great, admired. You are considered one of the world's geniuses. But you could not make a *second career*."[2] Ayers goads her to try again—to try to become famous a second time. Nash agrees to try to make another career, to show that what she had achieved she could achieve again, easily, by the force of her genius and personality alone.

But of course, she can't. Nash finds out the hard way that she cannot start over in Hollywood and make a second career that in any way achieves the levels of fame and fortune she had the first time around.

———

The stories of famous fossils rising to celebrity status are a bit like Rand's character Claire Nash. The stories of these seven fossils speak, in no small way, to the fragility of fame and the contingency of celebrity. Throughout the twentieth and twenty-first centuries, the discovery of fossil hominins has depended on a bit of luck and a bit of skill, especially concerning their cultural provenience. The fossils' narratives—their paths to celebrity—are one big argument for the power of historical contingency.

Stephen Jay Gould explored the power of contingency in explaining evolution writ large when he proposed rewinding the tape of evolution backward and replaying it. The question—would life evolve in the same way or in some very different manner?—is like asking if replaying the tape of evolution would make every species' evolutionary history as a "second career." Gould's metaphor, replaying a species' phylogeny, tells us that a species' evolutionary history is an unrepeatable series of events.

Gould delved further into the concept of contingency in his 1985 book *The Flamingo's Smile*. The flamingo, Gould informed readers, is an organism with an odd juxtaposition of beak shape and feeding behavior. Most birds feed by moving the bottom part of their beaks up and down. However, when flamingos dip their heads into water to feed, the relative positions of the beak halves change, meaning that the flamingo can't feed "normally" because its head is upside down. But flamingo beaks exhibit a particularly curious evolutionary characteristic: the beak as a mobile ball-and-socket joint, which allows the bird to change which part of the beak moves depending on what it is doing. Thus, if the bird is preening, the bottom beak moves because the bird is normal side up—but if the bird is feeding, then the top beak moves because the bird's head is upside down. According to Gould, the flamingo is a brilliant example of successfully inverting nature to live life upside down. The topsy-turvy, crisscrossing evolutionary history of flamingos means that they show successful beak adaptations, but, as Gould emphasized, the path to the flamingo's beak would have been completely unrepeatable. "Nature harbors a large suite of oddities so special that we scarce know how to predict," Gould concludes.[3]

For these seven fossils, their cultural histories—just like their evolutionary ones—are made up of twists and turns. These celebrities of the fossil record live as props, mascots, symbols, and avatars in stories of paleoanthropology's science, but they are also objects with their own cultural trajectories that are just as unique and unrepeatable as the evolution of the species they represent. After their discoveries, these fossils go this way and then that way—pushed by one scientist to explain a particular model of evolution and pulled by another to explain something else, held up as an exemplar of good science or derided as what happens when science goes wrong. Most important, though, the fossils are the sum of these stories—their celebrity is an artifact of accidents, happenstances of history, small but purposeful decisions by people that add up to bigger things.

These historical happenstances are set into motion only when the fossils are discovered, putting their discoverers squarely in the limelight. For these seven fossils in particular, their discoverers serve as advocates and interpreters over their professional careers, making them, for better or worse, those with the final say about the fossils. This status—a fossil's social keeper—offers a certain celebrity all its own, making the discoverers famous in their own right. In *The New Celebrity Scientists: Out of the Lab and into the Limelight*, sociologist Declan Fahy suggests that celebrity can have a positive power. "Celebrities personify and act as figureheads. . . . Celebrities with enduring popularity and prominence have a way of portraying the deep questions, tensions, and conflicts of their eras. Celebrities come to personify the culture and society of their particular time and place," he argues. "They help people make sense of the world."[4]

This creates an inexorable link between these scientists and the fossils they discover—they rise and fall as the *rota fortunae* of a fossil's acceptance turns throughout the decades. When the Taung Child and its species, *Australopithecus africanus*, were finally accepted as a legitimate human ancestor, Raymond Dart was welcomed back to the scientific establishment. The excitement of discovering a *Homo* ancestor at Zhoukoudian meant that Davidson Black and Johan Gunnar Andersson could enjoy institutional support to establish a serious lab in Beijing. Lucy made Donald Johanson's career. The popular success of Sediba whetted scientific and public interests in Lee Berger's Next Big Thing—excavations of *Homo naledi*, which are currently a raging success. To be a successful celebrity fossil, then, is to balance the ever wobbly triple point of media, commoditization, and representation.

Although celebrity scientists are often propped up by their fossil discoveries, celebrity fossils are the culmination of thousands of decisions made over and over and over. These decisions—about how to study and how to internalize the fossil—add up and show what the fossil's audiences value. A good story of discovery is what really sets a celebrity fossil on its initial trajectory of fame, and, certainly, great discovery stories give public audiences a way of identifying with the fossil and its discoverer. But, fundamentally, fossils are discovered by chance, as Winston Ayers might remind Claire Nash, and are an unrepeatable type of scientific find. Just like their histories.

Maybe other fossils should also be given a chance at fame, as Winston Ayers suggested should be the case for the thousands of aspiring Hollywood starlets that are not Claire Nash. Here I'm reminded of the question that my colleague posed about how to write about famous fossils: "How could you write a book about famous fossils and not write about these other important fossils???" There are, of course, other important fossils in paleoanthropology's history, but these fossils are simply not elevated to the same celebrity since they do not resonate in popular ways with audiences. For the famous fossils, they are cultural figures with personae and symbolism; the stories they tell are beyond merely a simplistic one of discovery, observation, and theory. The better we understand fossils' metastories, the more clearly we can consider how we think about the interplay between science, history, and popular culture.

Anthropologist Elizabeth Hallam poignantly suggests that bones— indeed, fossils—are particularly adept at complex life histories. "Bones [occupy] diverse post-mortem lives: trophies, souvenirs, sources of knowledge, things to possess and trade, deceased relatives, scientific data, once living persons. . . . Bones have been sensed in emotional terms, known in empirical ways, collected and displayed, deemed necessary to bury, exhume and rebury. They might be preserved or obliterated, are sometimes openly memorialized and at others concealed and lost."[5] Our celebrity skeletons move easily between emotional and empirical elements as their afterlives unfold. For some, that's seeing The Real Fossil in a museum exhibit and feeling the authenticity of the artifact; for others, it's recognizing the fossil hominin at the beginning of *Man of Steel* when Jor-El retrieves the codex. (It's based on Mrs. Ples.)

How we talk about these famous discoveries shows how we construct a fossil's life story—its scientific value as well as cultural cachet. The seven fossils in this book are scientific objects, of course, but they also speak to how we think about science and scientific discovery in popular culture. In fact, there is no single path to celebrity. There is only the value—need?—of it at the path's end. Perhaps fifty or a hundred years from now, some of the B-list fossils will be celebrities, but, for now, they're not.

"One knows the tale," Joseph Campbell wrote in *The Hero with a Thousand Faces*. "It has been told a thousand ways."[6] The stories, the narratives, the ever growing archives of meaning and cultural ephemera that are associated with these famous scientific discoveries have been told and retold—hundreds, if not thousands, of ways—forming the fossils' mythos and giving them lives beyond that of a static object. How we understand the fossils of human evolution—which is to say, how we understand the origins of ourselves—is an integral part of the fossils' cultural history. We, in our own contexts and encounters with these hominin fossils, are contributing to their life stories—and, even more interestingly, we are actively engaged in the writing of those stories. The fossils' stories are still unfolding.

I think back to that wintery June morning in Johannesburg years ago when Dr. Tobias introduced me and a dozen or so other undergrad fossil enthusiasts to the iconic Taung Child. Sure, he spoke about the fossil's anatomy and biology, the uniqueness of Taung's fossilized brain cast, and what that was able to tell us about human ancestors from three million years ago. He talked through the evolutionary importance of the Taung Child's species, *Australopithecus africanus*. He even described open research questions—questions for which the fossil is still, almost one hundred years after its discovery, considered crucial evidence for scientists working on current human origins inquires.

But, more important, his lecture also demonstrated that the fossil is completely and utterly imbued with historical and cultural meaning—meaning that he, himself, was actually creating by giving that very lecture, just as he had given that lecture hundreds of times before. His stories about his own adviser, Raymond Dart, and Dart's adventures with the "missing link" were—are!—just as much a part of the fossil's life story as its 3-D scans, its caliper measurements, and the hundreds of its casts that circulate among museums. The fossil's poetic fan fiction and its small wooden box, left in a London taxicab by Dart's wife—to say nothing of Tobias's ventriloquist act with the Taung Child—are significant parts of the fossil's history, chapters in the fossil's life. Seeing the fossil—first in its fossil vault and later as casts in museums—means that I have participated in the life history of the fossil, just like everyone else who has done the same thing.

"One can think of [the Taung Child] as beautiful, both in terms of its scientific importance," University of the Witwatersrand curator Bernhard Zipfel suggests, "and its aesthetic characteristics so reminiscent of a work of art. It evokes emotion in those who see it. I experience gooseflesh every time I carefully pick it up."[7] The story of the Taung Child, like all of these fossils from Lucy to Flo to the Old Man, is far from over. Each new story—each and every new scientific study, museum exhibit, and pop culture reference—opens up the next chapter in the fossils' lives.

Their futures are still being written.

ACKNOWLEDGMENTS

A book like *Seven Skeletons* draws from myriad different fields and perspectives—I am much obliged for the feedback, conversations, suggestions, support, and enthusiasm that so many colleagues, experts, and friends have offered over the course of this project: Justin Adams, Stacey Ake, Lee Berger, Jan Ebbestad, Kevin Egan, Jan Freedman, Yohannes Haile-Selassie, Ronald Harvey, John Hawks, Charles D. Heim, Charles J. D. Heim, Lindsay Hunter, David Jones, William Jungers, Jon Kalb, John Kappelman, Linda Kim, Scott Knowles, Robert Kruszynski, Tanya Kulik, Kevin Kuykendall, Siu Kwan Lam, Kristi Lewton, Christopher Manias, Elizabeth Marima, John Mead, Nancy Odegaard, Sven Ouzman, Tammy Peters, Julien Riel-Salvatore, Sara Schechner, Karolyn Schindler, Shuk On Sham, Amy Slaton, Francis Thackery, Dirk Van Tuerenhout, Kirsten Vannix, Milford Wolpoff, and Bernhard Zipfel.

Additionally, many institutions have been kind enough to help facilitate the book's research through interviews, access to archives, copies of publications, and/or financial support, including: *The Appendix*, Bone Clones, Natural History Museum (London), Pennoni Honors College (Drexel University), Science Photo and Science

Source Library, Smithsonian Institution Archives, University of Texas at Austin Libraries, University of Texas at Austin's Institute for Historical Studies, University of the Witwatersrand (Archives), and Uppsala's Museum of Evolution.

I am indebted to my agent, Geri Thoma, and editor, Melanie Tortoroli, for their interest in this project and their help in taking *Seven Skeletons* from "idea" to "book." Holly Zemsta was kind enough to share her thoughts and feedback on many early drafts. My parents have always been excited to "talk fossils," and I'm glad that these seven haven't worn out their welcome. I am also most grateful to Stan Seibert for his unwavering optimism and enthusiasm for this project.

NOTES

INTRODUCTION. Famous Fossils, Hidden Histories

1. Joni Brenner, Elizabeth Burroughs, and Karel Nel, *Life of Bone: Art Meets Science* (Johannesburg: Wits University Press, 2011), p. 84.
2. Daniel J. Boorstin, *The Image: A Guide to Pseudo-Events in America* (New York: Vintage, 2012), p. 61.
3. Samuel Alberti, ed., *The Afterlives of Animals: A Museum Menagerie* (Charlottesville: University of Virginia Press, 2011), p. 1.
4. Brenner, Burroughs, and Nel, *Life of Bone*, p. 12.

CHAPTER ONE. The Old Man of La Chapelle: The Patriarch of Paleo

1. Lynn Barber Cardiff, *The Heyday of Natural History* (New York: Doubleday, 1984); Peter Dear, *Revolutionizing the Sciences: European Knowledge and Its Ambitions, 1500–1700*, second ed. (Princeton, NJ: Princeton University Press, 2009).
2. J. C. Fuhlrott, "Teilen des menschlichen Skelettes im Neanderthal bei Hochtal," *Verhandlungen des Naturhistorischen Vereins der preussischen Rheinlande und Westphalens* 14 (1856), p. 50; H. Schaaffhausen, ibid., pp. 38–42 and 50–52.
3. Ian Tattersall, *The Last Neanderthal: The Rise, Success, and Mysterious Extinction of Our Closest Human Relatives*, revised ed. (New York: Basic Books, 1999), pp. 74–77.
4. Ibid.
5. Fuhlrott and Schaaffhausen, "Teilen des menschlichen Skelettes."
6. Thomas Henry Huxley, *Man's Place in Nature* (Ann Arbor: University of Michigan Press, 1959), p. 205.

7. Marianne Sommer, "Mirror, Mirror on the Wall: Neanderthal as Image and 'Distortion' in Early 20th-Century French Science and Press," *Social Studies of Science* 36, no. 2 (April 1, 2006), pp. 207–40.

8. Jean Bouyssonie, "La Sepulture Moustérienne de La Chapelle-aux-Saints," *Cosmos*, July 9, 1909, p. 11.

9. Marianne Sommer, *Bones and Ochre: The Curious Afterlife of the Red Lady of Paviland* (Cambridge, MA: Harvard University Press, 2007), p. 176.

10. Lydia Pyne, "Neanderthals in 3D: L'Homme de La Chapelle," *Public Domain Review*, February 11, 2015.

11. Marcellin Boule, *L'Homme Fossile de La Chapelle-aux-Saints* (Paris: Masson, 1911), p. 11.

12. Sommer, "Mirror, Mirror."

13. Richard Milner and Rhoda Knight Kalt, *Charles R. Knight: The Artist Who Saw Through Time* (New York: Harry N. Abrams, 2012).

14. Lydia V. Pyne and Stephen J. Pyne, *The Last Lost World: Ice Ages, Human Origins, and the Invention of the Pleistocene* (New York: Viking, 2012).

15. J. H. Rosny, *The Quest for Fire* (New York: Ballantine, 1982), p. 6.

16. Lydia Pyne, "Quests for Fire: Neanderthals and Science Fiction," *Appendix* 2, no. 3 (July 2014); Lydia Pyne, "Our Neanderthal Complex," *Nautilus* 24 (May 14, 2015).

17. Boule, *L'Homme Fossile*, p. 10.

18. "Human Skull from Fontéchevade, France: Abstract," *Nature*.

19. William L. Straus, Jr., and A. J. E. Cave, "Pathology and the Posture of Neanderthal Man," *Quarterly Review of Biology* 32, no. 4 (December 1, 1957), pp. 348–63.

20. Pamela Jane Smith, "Professor Dorothy A. E. Garrod: 'Small, Dark, and Alive!,'" *Bulletin of the History of Archaeology* 7, no. 1 (May 20, 1997).

21. C. Loring Brace et al., "The Fate of the 'Classic' Neanderthals: A Consideration of Hominid Catastrophism," *Current Anthropology* 5, no. 1 (February 1, 1964), pp. 3–43.

22. N. C. Tappen, "The Dentition of the 'Old Man' of La Chapelle-aux-Saints and Inferences Concerning Neanderthal Behavior," *American Journal of Physical Anthropology* 67, no. 1 (May 1, 1985), p. 43.

23. Ibid.

24. William Rendu et al., "Evidence Supporting an Intentional Neandertal Burial at La Chapelle-aux-Saints," *Proceedings of the National Academy of Sciences* 111, no. 1 (January 7, 2014); emphasis added.

25. J. Gurche, *Shaping Humanity: How Science, Art, and Imagination Help Us Understand Our Origins* (New Haven, CT: Yale University Press, 2013).

26. M. Boule, *Fossil Men: A Textbook of Human Palaeontology* (Oak Brook, IL: Dryden Press, 1957).

27. William Shakespeare, *The Tempest*, Act I, Scene 2, lines 296–98, 363–65.

28. Almudena Estalrrich and Antonio Rosas, "Handedness in Neandertals from the El Sidrón (Asturias, Spain): Evidence from Instrumental Striations with

Ontogenetic Inferences," *PLOS ONE* 8, no. 5 (May 6, 2013), e62797; L. V. Golovanova et al., "Mezmaiskaya Cave: A Neanderthal Occupation in the Northern Caucasus," *Current Anthropology* 40, no. 1 (February 1999), pp. 77–86; Julien Riel-Salvatore, "A Spatial Analysis of the Late Mousterian Levels of Riparo Bombrini (Balzi Rossi, Italy)," *Canadian Journal of Archaeology* 37, no. 1 (2013), pp. 70–92; Julien Riel-Salvatore, interview with author, September 24, 2014.

CHAPTER TWO. Piltdown: A Name Without a Fossil

1. Frank Spencer, *The Piltdown Papers, 1908–1955: The Correspondence and Other Documents Relating to the Piltdown Forgery* (New York: Natural History Museum Publications and Oxford University Press, 1990), p. 17.
2. Ibid.
3. On Sunday, December 22, 1912, a *New York Times* headline screamed, "Darwin Theory Is Proved True. English Scientists Say the Skull Found in Sussex Establishes Humans Descent from Apes. Bones Illustrate a Stage of Human Evolution Which Has Only Been Imagined Before."
4. Dawson and Smith Woodward, as quoted in Spencer, *Piltdown Papers*, p. 15.
5. Ibid., p. 16.
6. Ibid., p. 17.
7. Arthur Smith Woodward, *The Earliest Englishman* (London: Watts, 1948), pp. 9–10.
8. Ibid.
9. Spencer, *Piltdown Papers*, p. 20.
10. "The Piltdown Bones and 'Implements,'" *Nature* 174, no. 4419 (July 10, 1954), pp. 61–62.
11. William Boyd Dawkins, "The Geological Evidence of Britain as to the Antiquity of Man," *Geology Magazine* 2: 464–66 (1915).
12. Henry Fairfield Osborn, *Men of the Old Stone Age, Their Environment, Life and Art* (New York: C. Scribner's Sons, 1925), p. 130.
13. *A Guide to the Fossil Remains of Man in the Department of Geology and Palaeontology in the British Museum (Natural History)* (London: British Museum, 1918), p. 14.
14. Raf De Bont, "The Creation of Prehistoric Man: Aimé Rutot and the Eolith Controversy, 1900–1920," *Isis* 94, no. 4 (December 2003), pp. 604–30.
15. Grafton Elliot Smith, *The Evolution of Man: Essays* (London: Oxford University Press, H. Milford, 1927), as quoted in John Reader, *Missing Links: The Hunt for Earliest Man* (London: Penguin, 1981), p. 68.
16. Reader, *Missing Links*, p. 71.
17. Joseph Sidney Weiner, Kenneth Page Oakley, and Wilfrid Edward Le Gros Clark, *The Solution of the Piltdown Problem* (London: British Museum, 1953), p. 53.
18. Charles Blinderman, *The Piltdown Inquest* (Buffalo, NY: Prometheus Books, 1986), p. 66.
19. Weiner, Oakley, and Clark, *The Solution of the Piltdown Problem*, p. 53.
20. Karolyn Schindler, "Piltdown's Victims: Arthur Smith Woodward," *Evolve* 11 (2012), pp. 32–37.

21. F. J. M. Postlethwaite, "Letter to Editor," *The Times* (London), November 25, 1953.
22. Piltdown Collection, Natural History Museum, London.
23. N. P. Morris, "The Piltdown Story," June 1954, Piltdown Collection, Natural History Museum, London.
24. Blinderman, *Piltdown Inquest*, p. 79.
25. Rosemary Powers, "Memo to Dr. Oakley," April 28, 1967, Piltdown Misc., Piltdown Collection, Natural History Museum, London.
26. Kenneth L. Feder, *Frauds, Myths, and Mysteries: Science and Pseudoscience in Archaeology* (Boston: McGraw-Hill Mayfield, 2001), p. 55.
27. Claude Levi-Strauss, *Myth and Meaning: Cracking the Code of Culture* (New York: Schocken, 1978), pp. 40–41.
28. Schindler, "Piltdown's Victims," p. 37.

CHAPTER THREE. The Taung Child: The Rise of a Folk Hero

1. Raymond A. Dart with Dennis Craig, *Adventures with the Missing Link* (New York: Harper and Brothers, 1959), pp. 6–7.
2. As quoted in Roger Lewin, *Bones of Contention: Controversies in the Search for Human Origins*, second edition (Chicago: University of Chicago Press, 1997), p. 50.
3. Dart and Craig, *Adventures with the Missing Link*, p. 4.
4. Ibid., pp. 6–7.
5. Raymond Dart, "*Australopithecus Africanus*: The Man-Ape of South Africa," *Nature* 115, no. 2884 (1925), pp. 195–99; Reader, *Missing Links*, p. 82.
6. Dart and Craig, *Adventures with the Missing Link*, p. 10.
7. Dart, "*Australopithecus*"; emphasis in original.
8. Ibid., pp. 198–99.
9. Dart and Craig, *Adventures with the Missing Link*, pp. 6–7.
10. Letter from F. O. Barlow, dated October 17, 1928, Raymond Dart Archive, University of the Witwatersrand.
11. Anne Clendinning, "On the British Empire Exhibition, 1924–25," Branch Collective.
12. Letter (Published) from the Exhibition Commissioner, dated July 9, 1925, correspondence in the Raymond Dart Archive, University of the Witwatersrand.
13. Raymond Dart Archive, University of the Witwatersrand.
14. Letter (Published) from the Exhibition Commissioner, dated July 9, 1925, correspondence in the Raymond Dart Archive, University of the Witwatersrand.
15. Raymond Dart Archive, University of the Witwatersrand; Arthur Keith, "Letter to Editor," *Nature* 116 (September 26, 1925), pp. 462–63.
16. Raymond Dart Archive, University of the Witwatersrand.
17. Letter from Joseph Liddle, dated May 3, 1930, correspondence in the Raymond Dart Archive, University of the Witwatersrand.

18. Dart and Craig, *Adventures with the Missing Link,* as contextualized by Reader, *Missing Links.*
19. Manisha R. Dayal et al., "The History and Composition of the Raymond A. Dart Collection of Human Skeletons at the University of the Witwatersrand, Johannesburg, South Africa," *American Journal of Physical Anthropology* 140, no. 2 (2009), pp. 324–35.
20. Dart and Craig, *Adventures with the Missing Link*; Reader, *Missing Links.*
21. Reader, *Missing Links.*
22. Lewin, *Bones of Contention,* p. 47.
23. Raymond Dart Archive, University of the Witwatersrand.
24. Ibid.
25. C. K. Brain et al., "New Evidence of Early Hominids, Their Culture and Environment, from Swartkrans Cave, South Africa," *South African Journal of Science* 84 (1988), pp. 828–35.
26. Charles K. Brain et al., *Staatsmuseum 100: National Cultural History Museum, Museum of the Geological Survey, Transvaal Museum,* National Cultural History Museum, 1992; Tersia Perregil, Ditsong Museum archivist, e-mail interview with author, January 2014.
27. Lydia Pyne, "Ditsong's Dioramas: Putting a Body on a Fossil and a Fossil in a Narrative," *Appendix* 2, no. 2 (April 2014).
28. Other writers and artists contributed to the publication *Life of Bone: Art Meets Science*; Brenner, Burroughs, and Nel, *Life of Bone,* p. 9.
29. Brenner, Burroughs, and Nel, *Life of Bone.*
30. Kristi Lewton, e-mail and phone interview with author, February 28, 2014, and March 3, 2014.
31. Brenner, Burroughs, and Nel, *Life of Bone,* p. 3.
32. Lee Berger, interview with author, June 27, 2013, University of the Witwatersrand.
33. Kristi Lewton, e-mail and phone interview with author, February 28, 2014, and March 3, 2014.

CHAPTER FOUR. Peking Man: A Curious Case of Paleo-Noir

1. Anneli Waara, "Unique Tooth Reveals Details of the Peking Man's Life," Uppsala University; Jan Petter Myklebust, "Tooth of 'Peking Man' Found Again After 90 Years," University World News, March 20, 2015.
2. Lanpo Jia and Weiwen Huang, *The Story of Peking Man: From Archaeology to Mystery* (Oxford: Oxford University Press, 1990), p. 10.
3. Peter C. Kjaergaard, "The Missing Links Expeditions—Or How the Peking Man Was Not Found," *Endeavour* 36, no. 3 (September 2012), pp. 97–105.
4. Ibid., p. 98.
5. Johan Gunnar Andersson, *Children of the Yellow Earth: Studies in Prehistoric China,* reprint (Cambridge, MA: MIT Press, 1973).
6. Kjaergaard, "Missing Links Expeditions," p. 97.
7. Jia and Huang, *Story of Peking Man,* p. 20.
8. Ibid., p. 49.

9. Ibid., pp. 63–64.
10. Ibid., pp. 64–65.
11. Ibid., p. 65.
12. Ibid., p. 66.
13. Hsiao-pie Yen, "Constructing the Chinese: Paleoanthropology and Anthropology in the Chinese Frontier, 1920–1950," doctoral dissertation, Harvard University, 2012.
14. Rockefeller Foundation, RG 1.2, Series 601D (China), Box 1, Folder 4: China, PUMC: Davidson Black (courtesy of Christopher Manias).
15. Chris Manias, e-mail interview with author, May 20, 2015.
16. Grace Yen Shen, *Unearthing the Nation: Modern Geology and Nationalism in Republican China* (Chicago: University of Chicago Press, 2013), p. 5.
17. Chris Manias, e-mail interview with author, May 20, 2015.
18. Jia and Huang, *Story of Peking Man*, p. 175, as quoting Ruth Moore.
19. Christopher G. Janus and William Brashler, *The Search for Peking Man* (New York: Macmillan, 1975).
20. "Financier Is Charged with Fraud in Search for Bones of Peking Man," Reuters, February 26, 1981; Stephen Miller, "Colorful Chicagoan's Biggest Stunt, Detective Mission to Find Peking Man, Led to Fraud Plea," *Wall Street Journal*, February 28, 2009.
21. Miller, "Colorful Chicagoan's Biggest Stunt."
22. Jane Hooker, "Letter from China: The Search for Peking Man," *Archaeology*, March/April 2006.
23. Lydia Pyne, "To Russia, with Love," *Appendix* 2, no. 4 (October 2014).
24. Raymond Dart Archive, University of the Witwatersrand.
25. Amir D. Aczel, *The Jesuit and the Skull: Teilhard de Chardin, Evolution, and the Search for Peking Man* (New York: Riverhead, 2007), p. 154.
26. "Reproducing Our Ancestors," *Expedition Magazine* 29, no.1 (March 1987); www.penn.museum/sites/expedition/reproducing-our-ancestors.
27. Ibid.
28. Jia and Huang, *Story of Peking Man*, pp. 174–75; Harry L. Shapiro, *Peking Man: The Discovery, Disappearance and Mystery of a Priceless Scientific Treasure* (New York: Simon & Schuster, 1974), p. 30.
29. Yen, "Constructing the Chinese," pp. 10–11.
30. Waara, "Unique Tooth Reveals Details of the Peking Man's Life."

CHAPTER FIVE. The Ascension of an Icon: Lucy in the Sky

1. Donald Johanson and Maitland Edey, *Lucy: The Beginnings of Humankind* (New York: Simon & Schuster, 1981).
2. "Ancient *Homo Sapiens* Found in Central Afar," *Ethiopian Herald*, October 26, 1974.
3. Johanson and Edey, *Lucy*, p. 18.
4. Lauren E. Bohn, "Q&A: 'Lucy' Discoverer Donald C. Johanson," *Time*, March 4, 2009.

5. "In Central Afar: Most Complete Remains of Man Discovered," *Ethiopian Herald*, December 21, 1974.

6. Ibid.

7. Jon E. Kalb, *Adventures in the Bone Trade: The Race to Discover Human Ancestors in Ethiopia's Afar Depression* (New York: Copernicus, 2001), pp. 150–51.

8. Ibid.

9. D. C. Johanson and M. Taieb, "Plio-Pleistocene Hominid Discoveries in Hadar, Ethiopia," *Nature* 260, no. 5549 (March 25, 1976), pp. 293–97.

10. Ibid.

11. Lewin, *Bones of Contention*, p. 271.

12. "Forty Years After Lucy's Ethiopia Discovery: A Conversation with Donald Johanson," *Tadias*, November 24, 2014.

13. Lewin, *Bones of Contention*, p. 270.

14. Richard Brilliant, *Portraiture* (London: Reaktion Books, 2003), p. 8.

15. Ibid., p. 61.

16. Pyne, "Ditsong's Dioramas."

17. Ann Gibbons, "Lucy's Tour Abroad Sparks Protests," *Science* 314, no. 5799 (October 27, 2006), pp. 574–75.

18. Ibid.

19. Ibid.

20. Dirk Van Tuerenhout, interview with author, November 15, 2012, and May 12, 2015.

21. Ibid.

22. Ibid.

23. Juliet Eilperin, "In Ethiopia, Both Obama and Ancient Fossils Get a Motorcade," *Washington Post*, July 27, 2015.

24. William Yardley, "They Didn't Love Lucy," *New York Times*, March 13, 2009.

25. Nancy Odegaard, phone interview with author, June 25, 2015.

26. Ibid.

27. Ronald Harvey, phone interview with author, June 26, 2015.

28. Ibid.

29. Eilperin, "In Ethiopia, Both Obama and Ancient Fossils."

30. Nancy Odegaard, phone interview with author, June 25, 2015.

31. Donald Johanson and James Shreeve, *Lucy's Child: The Discovery of a Human Ancestor* (New York: Harper Perennial, 1990).

32. E. F. K. Koerner, *Ferdinand de Saussure: Origin and Development of His Linguistic Thought in Western Studies of Language: A Contribution to the History and Theory of Linguistics, Schriften zur Linguistik* 7 (Braunschweig: Vieweg, 1973); Carol Sanders, ed., *The Cambridge Companion to Saussure* (New York: Cambridge University Press, 2004).

33. Ronald Harvey, phone interview with author, June 26, 2015.

34. Kristi Lewton, e-mail and phone interview with author, February 28, 2014, and March 3, 2014.

35. Bone Clones, e-mail interview with author, May 14, 2015.

CHAPTER SIX. The Precious: Flo's Life as a Hobbit

1. Ewen Callaway, "The Discovery of *Homo Floresiensis*: Tales of the Hobbit," *Nature* 514, no. 7523 (October 23, 2014), pp. 422–26.

2. M. J. Morwood and Penny Van Oosterzee, *A New Human: The Startling Discovery and Strange Story of the "Hobbits" of Flores, Indonesia* (New York: Smithsonian Books/Collins, 2007), p. 27.

3. Ibid., p. 31.

4. Ibid., p. 85.

5. Callaway, "Discovery of *Homo Floresiensis*."

6. Tabitha Powledge, "Skullduggery: The Discovery of an Unusual Human Skeleton Has Broad Implications," *EMBO Reports* 6 (2005), pp. 609–12.

7. Callaway, "Discovery of *Homo Floresiensis*."

8. Ibid.

9. Ibid.

10. Michael Hopkin, "Wrist Bones Bolster Hobbit Status," *Nature News*, September 20, 2007; Matthew W. Tocheri et al., "The Primitive Wrist of *Homo Floresiensis* and Its Implications for Hominin Evolution," *Science* 317, no. 5845 (September 21, 2007), pp. 1743–45.

11. Callaway, "Discovery of *Homo Floresiensis*."

12. "Rude Palaeoanthropology," *Nature* 442, no. 7106 (August 31, 2006), p. 957.

13. Quotes from Marta Mirazon Lahr and Robert Foley, "Palaeoanthropology: Human Evolution Writ Small," *Nature* 431 (October 28, 2004), p. 1043; Michael Hopkin, "The Flores Find," *Nature News* (October 27, 2004).

14. Mirazon Lahr and Foley, "Palaeoanthropology," pp. 1043–44.

15. Lachlan Williams, "Academia Is 'Bitchy': Fight Erupts over 'Hobbit' Fossil," 9 Stories, NineMSN, September 23, 2014; Maciej Henneberg et al., "Evolved Developmental Homeostasis Disturbed in LB1 from Flores, Indonesia, Denotes Down Syndrome and Not Diagnostic Traits of the Invalid Species *Homo Floresiensis*," *Proceedings of the National Academy of Sciences* 111, no. 33 (August 4, 2014), 201407382.

16. Callaway, "Discovery of *Homo Floresiensis*."

17. Rex Dalton, "Little Lady of Flores Forces Rethink of Human Evolution," *Nature* 431, no. 1029 (October 28, 2004).

18. Gregory Forth, "Hominids, Hairy Hominoids and the Science of Humanity," *Anthropology Today* 21, no. 3 (June 1, 2005): pp. 13–17.

19. John Gurche, *Shaping Humanity: How Science, Art, and Imagination Help Us Understand Our Origins* (New Haven, CT: Yale University Press, 2013), pp. 270–71.

20. Dean Falk, *The Fossil Chronicles: How Two Controversial Discoveries Changed Our View of Human Evolution* (Oakland: University of California Press, 2012), p. 78.

21. Callaway, "Discovery of *Homo Floresiensis*."

22. Ibid.

23. Morwood and Oosterzee, *A New Human*, p. xii.

CHAPTER SEVEN. Sediba: TBD (To Be Determined)

1. Celia W. Dugger and John Noble Wilford, "New Hominid Species Discovered in South Africa," *New York Times*, April 8, 2010.
2. Lee R. Berger, *Working and Guiding in the Cradle of Humankind* (Johannesburg: Prime Origins, 2005).
3. Rex Dalton, "Africa's Next Top Hominid," *Nature News*, June 21, 2010.
4. Yohannes Haile-Selassie et al., "An Early *Australopithecus Afarensis* Postcranium from Woranso-Mille, Ethiopia," *Proceedings of the National Academy of Sciences* 107, no. 27 (July 6, 2010), pp. 12121–26.
5. Dalton, "Africa's Next Top Hominid."
6. Philip L. Reno and C. Owen Lovejoy, "From Lucy to Kadanuumuu: Balanced Analyses of *Australopithecus Afarensis* Assemblages Confirm Only Moderate Skeletal Dimorphism," *PeerJ* 3 (April 28, 2015), p. e925.
7. "Wits Scientists Reveal New Species of Hominid," University of the Witwatersrand, April 8, 2010.
8. Ibid.
9. Berger, *Working and Guiding*.
10. "Wits Scientists Reveal New Species of Hominid," University of the Witwatersrand.
11. Kate Wong, "Is *Australopithecus Sediba* the Most Important Human Ancestor Discovery Ever?," *Scientific American*, April 24, 2013.
12. Fred Spoor, "Palaeoanthropology: Malapa and the Genus *Homo*," *Nature* 478, no. 7367 (October 6, 2011), pp. 44–45.
13. "Wits Scientists Reveal New Species of Hominid," University of the Witwatersrand.
14. Ker Than, "Surprise Human-Ancestor Find—Key Fossils Hidden in Lab Rock," *National Geographic News*, July 14, 2012.
15. "Rising Star Empire Cave 2014 Annual Report," SAHRA.
16. Wong, "Is *Australopithecus Sediba* the Most Important Human Ancestor Discovery Ever?"
17. Yohannes Haile-Selassie et al., "New Species from Ethiopia Further Expands Middle Pliocene Hominin Diversity," *Nature* 521, no. 7553 (May 28, 2015), pp. 483–88.
18. Kate Wong, "Could a Renewed Push for Access to Fossil Data Finally Topple Paleoanthropology's Culture of Secrecy?," *Scientific American*, May 8, 2012.
19. Ibid.

AFTERWORD. *O Fortuna!*: A Bit of Luck, a Bit of Skill

1. Ayn Rand and Leonard Peikoff, *The Early Ayn Rand: A Selection from Her Unpublished Fiction* (New York: New American Library, 1984), p. 89.
2. Ibid., pp. 93–94.
3. Stephen Jay Gould, *The Flamingo's Smile: Reflections in Natural History* (New York: W. W. Norton, 1985), p. 26.

4. Declan Fahy, *The New Celebrity Scientists: Out of the Lab and into the Limelight* (Lanham, MD: Rowman & Littlefield, 2015), p. 7.

5. Elizabeth Hallam, "Articulating Bones: An Epilogue," *Journal of Material Culture* 15, no. 4 (December 1, 2010), pp. 465–66.

6. Joseph Campbell, *The Hero with a Thousand Faces*, reprint (San Francisco: New World Library, 2008), p. 334.

7. Brenner, Burroughs, and Nel, *Life of Bone*, p. 3.

BIBLIOGRAPHY

Aczel, Amir D. *The Jesuit and the Skull: Teilhard de Chardin, Evolution, and the Search for Peking Man*. New York: Riverhead, 2007.

Alberti, Samuel, ed. *The Afterlives of Animals: A Museum Menagerie*. Charlottesville: University of Virginia Press, 2011.

"Ancient *Homo Sapiens* Found in Central Afar." *Ethiopian Herald*. October 26, 1974.

Andersson, Johan Gunnar. *Children of the Yellow Earth: Studies in Prehistoric China*, reprint. Cambridge, MA: MIT Press, 1973 [1934].

Barlow, F. O. Letter dated October 17, 1928. Raymond Dart Archive, University of the Witwatersrand.

Berger, Lee R. *Working and Guiding in the Cradle of Humankind*. Johannesburg: Prime Origins, 2005.

———. Interview by Lydia Pyne, in person, June 27, 2013.

Blinderman, Charles. *The Piltdown Inquest*. Buffalo, NY: Prometheus, 1986.

Bohn, Lauren E. "Q&A: 'Lucy' Discoverer Donald C. Johanson." *Time*, March 4, 2009. http://content.time.com/time/health/article/0,8599,1882969,00.html.

Bone Clones. Interview by Lydia Pyne, e-mail, May 14, 2015.

Boorstin, Daniel J. *The Image: A Guide to Pseudo-Events in America*, reprint. New York: Vintage, 2012 [1961].

Boule, Marcellin. *Fossil Men: A Textbook of Human Palaeontology*. Oak Brook, IL: Dryden Press, 1957.

———. *L'Homme Fossile de La Chapelle-aux-Saints*. Paris: Masson, 1911.

Bouyssonie, Jean. "La Sepulture Moustérienne de La Chapelle-aux-Saints." *Cosmos*, July 9, 1909.

Brace, C. Loring, et al. "The Fate of the 'Classic' Neanderthals: A Consideration of Hominid Catastrophism." *Current Anthropology* 5, no. 1 (February 1, 1964): 3–43.

Brain, Charles Kimberlin, National Cultural History Museum (South Africa), Geological Survey Museum (South Africa), and Transvaal Museum. *Staatsmuseum 100: National Cultural History Museum, Museum of the Geological Survey, Transvaal Museum*. National Museum of Cultural History, 1992.

Brain, C. K., C. S. Chrucher, J. D. Clark, F. E. Grine, P. Shipman, R. L. Susman, A. Turner, and V. Watson. "New Evidence of Early Hominids, Their Culture and Environment, from Swartkrans Cave, South Africa." *South African Journal of Science* 84 (1988): 828–35.

Brenner, Joni, Elizabeth Burroughs, and Karel Nel. *Life of Bone: Art Meets Science*. Johannesburg: Wits University Press, 2011.

Brilliant, Richard. *Portraiture*. London: Reaktion Books, 2003.

Callaway, Ewen. "The Discovery of *Homo Floresiensis*: Tales of the Hobbit." *Nature* 514, no. 7523 (October 23, 2014): 422–26. doi:10.1038/514422a.

Campbell, Joseph. *The Hero with a Thousand Faces*, reprint. San Francisco: New World Library, 2008 [1949].

Cardiff, Lynn Barber. *The Heyday of Natural History*. New York: Doubleday, 1984.

Carlisle, Ronald C., and Michael I. Siegel. "Some Problems in the Interpretation of Neanderthal Speech Capabilities: A Reply to Lieberman." *American Anthropologist* 76, no. 2 (June 1, 1974): 319–22. doi:10.1525/aa.1974.76.2.02a00050.

Chojnacki, Stanislaw. *Ethiopian Icons: Catalogue of the Collection of the Institute of Ethiopian Studies, Addis Ababa University*. Milan: Skira, 2000.

Clendinning, Anne. "On the British Empire Exhibition, 1924–25." Branch Collective. www.branchcollective.org/?ps_articles=anne-clendinning-on-the-british-empire-exhibition-1924-25.

Dalton, Rex. "Africa's Next Top Hominid." *Nature News*, June 21, 2010. doi:10.1038/news.2010.305.

———. "Little Lady of Flores Forces Rethink of Human Evolution." *Nature* 431, no. 1029 (October 28, 2004). doi:10.1038/4311029a.

Dart, Raymond. "*Australopithecus Africanus*: The Man-Ape of South Africa." *Nature* 115, no. 2884 (1925): 195–99. doi:10.1038/115195a0.

Dart, Raymond A., with Dennis Craig. *Adventures with the Missing Link*. New York: Harper and Brothers, 1959.

Dawkins, William Boyd. *Early Man in Britain and His Place in the Tertiary Period*. London: Macmillan, 1880.

———. "The Geological Evidence in Britain as to the Antiquity of Man. *Geology Magazine* 2: 464–66 (1915).

Dayal, Manisha R., Anthony D. T. Kegley, Goran Štrkalj, Mubarak A. Bidmos, and Kevin L. Kuykendall. "The History and Composition of the Raymond A. Dart Collection of Human Skeletons at the University of the Witwatersrand, Johannesburg, South Africa." *American Journal of Physical Anthropology* 140, no. 2 (2009): 324–35. doi:10.1002/ajpa.21072.

Dear, Peter. *Revolutionizing the Sciences: European Knowledge and Its Ambitions, 1500–1700*, second ed. Princeton, NJ: Princeton University Press, 2009.

De Bont, Raf. "The Creation of Prehistoric Man: Aimé Rutot and the Eolith Controversy, 1900–1920." *Isis* 94, no. 4 (December 2003): 604–30. doi:10.1086/386384.

Dugger, Celia W., and John Noble Wilford. "New Hominid Species Discovered in South Africa." *New York Times*, April 8, 2010. www.nytimes.com/2010/04/09/science/09fossil.html.

Eilperin, Juliet. "In Ethiopia, Both Obama and Ancient Fossils Get a Motorcade." *Washington Post*, July 27, 2015. www.washingtonpost.com/blogs/post-politics/wp/2015/07/27/in-ethiopia-both-obama-and-ancient-fossils-get-a-motorcade/.

Estalrrich, Almudena, and Antonio Rosas. "Handedness in Neandertals from the El Sidrón (Asturias, Spain): Evidence from Instrumental Striations with Ontogenetic Inferences." *PLOS ONE* 8, no. 5 (May 6, 2013): e62797. doi:10.1371/journal.pone.0062797.

Fahy, Declan. *The New Celebrity Scientists: Out of the Lab and Into the Limelight.* Lanham, MD: Rowman & Littlefield, 2015.

Falk, Dean. *The Fossil Chronicles: How Two Controversial Discoveries Changed Our View of Human Evolution.* Oakland: University of California Press, 2012.

Feder, Kenneth L. *Frauds, Myths, and Mysteries: Science and Pseudoscience in Archaeology.* Boston: McGraw-Hill Mayfield, 2001.

"Financier Is Charged with Fraud in Search for Bones of Peking Man." Reuters, February 26, 1981. www.nytimes.com/1981/02/26/us/financier-is-charged-with-fraud-in-search-for-bones-of-peking-man.html.

Fiskesjö, Magnus. *China Before China: Johan Gunnar Andersson, Ding Wenjiang, and the Discovery of China's Prehistory.* Stockholm: Museum of Far Eastern Antiquities, 2004.

Forth, Gregory. "Hominids, Hairy Hominoids and the Science of Humanity." *Anthropology Today* 21, no. 3 (June 1, 2005): 13–17. doi:10.1111/j.0268-540X.2005.00353.x.

"Forty Years After Lucy's Ethiopia Discovery: A Conversation with Donald Johanson." *Tadias*, November 24, 2014. www.tadias.com/11/24/2014/forty-years-after-lucys-ethiopia-discovery-a-conversation-with-donald-johanson/.

Fuhlrott, J. C. "Teilen des menschlichen Skelettes im Neanderthal bei Hochtal." *Verhandlungen des Naturhistorischen Vereins der preussischen Rheinlande und Westphalens* 14 (1856): 50.

Gibbons, Ann. "Lucy's Tour Abroad Sparks Protests." *Science* 314, no. 5799 (October 27, 2006): 574–75. doi:10.1126/science.314.5799.574.

Golovanova, L. V., John F. Hoffecker, V. M. Kharitonov, and G. P. Romanova. "Mezmaiskaya Cave: A Neanderthal Occupation in the Northern Caucasus." *Current Anthropology* 40, no. 1 (February 1999): 77–86. doi:10.1086/515805.

Gould, Stephen Jay. *The Flamingo's Smile: Reflections in Natural History.* New York: W. W. Norton, 1987.

A Guide to the Fossil Remains of Man in the Department of Geology and Palaeontology in the British Museum (Natural History). British Museum (Natural History), Department of Geology. London: Trustees of the British Museum, 1918.

Gurche, John. *Shaping Humanity: How Science, Art, and Imagination Help Us Understand Our Origins*. New Haven, CT: Yale University Press, 2013.

Haile-Selassie, Yohannes, Luis Gibert, Stephanie M. Melillo, Timothy M. Ryan, Mulugeta Alene, Alan Deino, Naomi E. Levin, Gary Scott, and Beverly Z. Saylor. "New Species from Ethiopia Further Expands Middle Pliocene Hominin Diversity." *Nature* 521, no. 7553 (May 28, 2015): 483–88. doi:10.1038/nature14448.

Haile-Selassie, Yohannes, Bruce M. Latimer, Mulugeta Alene, Alan L. Deino, Luis Gibert, Stephanie M. Melillo, Beverly Z. Saylor, Gary R. Scott, and C. Owen Lovejoy. "An Early *Australopithecus Afarensis* Postcranium from Woranso-Mille, Ethiopia." *Proceedings of the National Academy of Sciences* 107, no. 27 (July 6, 2010): 12121–26. doi:10.1073/pnas.1004527107.

Hallam, Elizabeth. "Articulating Bones: An Epilogue." *Journal of Material Culture* 15, no. 4 (December 1, 2010): 465–92. doi:10.1177/1359183510382963.

Harvey, Ronald. Interview by Lydia Pyne, phone, June 26, 2015.

Henneberg, Maciej, Robert B. Eckhardt, Sakdapong Chavanaves, and Kenneth J. Hsü. "Evolved Developmental Homeostasis Disturbed in LB1 from Flores, Indonesia, Denotes Down Syndrome and Not Diagnostic Traits of the Invalid Species *Homo Floresiensis*." *Proceedings of the National Academy of Sciences* 111, no. 33 (August 4, 2014): 201407382. doi:10.1073/pnas.1407382111.

Henze, Paul B. *Layers of Time: A History of Ethiopia*. New York: St. Martin's, 2000.

Hooker, Jane. "Letter from China: The Search for Peking Man." *Archaeology*, March/April 2006. http://archive.archaeology.org/0603/abstracts/letter.html.

Hopkin, Michael. "The Flores Find." *Nature News*, October 27, 2004. doi:10.1038/news041025-4.

———. "Wrist Bones Bolster Hobbit Status." *Nature News*, September 20, 2007. doi:10.1038/news070917-8.

"Human Skull from Fontéchevade, France: Abstract." Nature. www.nature.com/nature/journal/v163/n4142/abs/163435b0.html.

Huxley, Thomas Henry. *Man's Place in Nature*. Ann Arbor: University of Michigan Press, 1959.

"In Central Afar: Most Complete Remains of Man Discovered." *Ethiopian Herald*. December 21, 1974.

Janus, Christopher G., and William Brashler. *The Search for Peking Man*. New York: Macmillan, 1975.

Jia, Lanpo, and Weiwen Huang. *The Story of Peking Man: From Archaeology to Mystery*. Translated by Yin Zhiqi. Oxford: Oxford University Press, 1990.

Johanson, D. C., and M. Taieb. "Plio-Pleistocene Hominid Discoveries in Hadar, Ethiopia." *Nature* 260, no. 5549 (March 25, 1976): 293–97. doi:10.1038/260293a0.

Johanson, Donald, and Maitland Edey. *Lucy: The Beginnings of Humankind*. New York: Simon & Schuster, 1981.

Johanson, Donald, and James Shreeve. *Lucy's Child: The Discovery of a Human Ancestor*. New York: Harper Perennial, 1990.

Johanson, Dr. Don~¹ᴶ, and Kate Wong. *Lucy's Legacy: The Quest for Human Origins.* 1 ed. N̶ w York: Broadway Books, 2010.

Kalb, Jon E. *Adventures in the Bone Trade: The Race to Discover Human Ancestors in Ethiopia's Afar Depression.* New York: Copernicus, 2001.

Keith, Arthur. "Letter to Editor." *Nature* 116 (September 26, 1925): 462–63.

Kjaergaard, Peter C. "The Missing Links Expeditions—Or How the Peking Man Was Not Found." *Endeavour* 36, no. 3 (September 2012): 97–105. doi:10.1016/j.endeavour.2012.01.002.

Koerner, E. F. K. *Ferdinand de Saussure: Origin and Development of His Linguistic Thought in Western Studies of Language: A Contribution to the History and Theory of Linguistics. Schriften zur Linguistik* 7. Braunschweig: Vieweg, 1973.

Letter (Published) from the Exhibition Commissioner, dated July 9, 1925. Raymond Dart Archive, University of the Witwatersrand.

Levi-Strauss, Claude. *Myth and Meaning: Cracking the Code of Culture.* New York: Schocken, 1978.

Lewin, Roger. *Bones of Contention: Controversies in the Search for Human Origins,* second ed. Chicago: University of Chicago Press, 1997 [1986].

Lewton, Kristi. Interview by Lydia Pyne, e-mail, February 28, 2014; phone, March 3, 2014.

Liddle, Joseph. Letter dated May 3, 1930. Raymond Dart Archive, University of the Witwatersrand.

Lieberman, Philip, and Edmund S. Crelin. "On the Speech of Neanderthal Man." *Linguistic Inquiry* 2, no. 2 (April 1, 1971): 203–22.

Manias, Christopher. Interview by Lydia Pyne, e-mail, May 20, 2015.

Miller, Stephen. "Colorful Chicagoan's Biggest Stunt, Detective Mission to Find Peking Man, Led to Fraud Plea." *Wall Street Journal,* February 28, 2009. www.wsj.com/articles/SB123579056359499267.

Milner, Richard, and Rhoda Knight Kalt. *Charles R. Knight: The Artist Who Saw Through Time.* New York: Harry N. Abrams, 2012.

Mirazon Lahr, Marta, and Robert Foley. "Palaeoanthropology: Human Evolution Writ Small." *Nature* 431, no. 7012 (October 28, 2004): 1043–44. doi:10.1038/4311043a.

Morris, N. P. "The Piltdown Story." June 1954. Piltdown Collection, Natural History Museum, London.

Morwood, M. J., and Penny Van Oosterzee. *A New Human: The Startling Discovery and Strange Story of the "Hobbits" of Flores, Indonesia.* New York: Smithsonian Books/Collins, 2007.

Myklebust, Jan Petter. "Tooth of 'Peking Man' Found Again After 90 Years." *University World News,* March 20, 2015. www.universityworldnews.com/article.php?story=20150320082920613.

Odegaard, Nancy. Interview by Lydia Pyne, phone, June 25, 2015.

Osborn, Henry Fairfield. *Men of the Old Stone Age, Their Environment, Life and Art.* New York: C. Scribner's Sons, 1925.

Perregil, Tersia, Ditsong Museum. Interview by Lydia Pyne, e-mail, January 2014.

"The Piltdown Bones and 'Implements.'" *Nature* 174, no. 4419 (July 10, 1954): 61–62. doi:10.1038/174061a0.

Postlethwaite, F. J. M. "Letter to Editor." *The Times*, November 25, 1953.

Powers, Rosemary. "Memo to Dr. Oakley." April 28, 1967. Piltdown Misc., Piltdown Collection, Natural History Museum, London.

Powledge, Tabitha. "Skullduggery: The Discovery of an Unusual Human Skeleton Has Broad Implications." *EMBO Reports* 6 (2005): 609–12.

Pyne, Lydia. "Ditsong's Dioramas: Putting a Body on a Fossil and a Fossil in a Narrative." *Appendix* 2, no. 2 (April 2014). http://theappendix.net/issues/2014/4/ditsongs-dioramas-putting-a-body-on-a-fossil-and-a-fossil-in-a-narrative.

———. "Neanderthals in 3D: L'Homme de La Chapelle." *Public Domain Review*, February 11, 2015. http://publicdomainreview.org/2015/02/11/neanderthals-in-3d-lhomme-de-la-chapelle/.

———. "Our Neanderthal Complex." *Nautilus* 24 (May 14, 2015).

———. "Quests for Fire: Neanderthals and Science Fiction." *Appendix* 2, no. 3 (July 2014). http://theappendix.net/issues/2014/7/quests-for-fire-neanderthals-and-science-fiction.

———. "To Russia, with Love." *Appendix* 2, no. 4 (October 2014). http://theappendix.net/issues/2014/10/to-russia-with-love.

Pyne, Lydia V., and Stephen J. Pyne. *The Last Lost World: Ice Ages, Human Origins, and the Invention of the Pleistocene*. New York: Viking, 2012.

Rand, Ayn, and Leonard Peikoff. *The Early Ayn Rand: A Selection from Her Unpublished Fiction*. New York: New American Library, 1984.

Reader, John. *Missing Links: The Hunt for Earliest Man*. London: Penguin, 1981.

Rendu, William, et al. "Evidence Supporting an Intentional Neandertal Burial at La Chapelle-aux-Saints." *Proceedings of the National Academy of Sciences* 111, no. 1 (January 7, 2014): 201316780. doi:10.1073/pnas.1316780110.

Reno, Philip L., and C. Owen Lovejoy. "From Lucy to Kadanuumuu: Balanced Analyses of *Australopithecus Afarensis* Assemblages Confirm Only Moderate Skeletal Dimorphism." *PeerJ* 3 (April 28, 2015): e925. doi:10.7717/peerj.925.

"Reproducing Our Ancestors." *Expedition Magazine*. www.penn.museum/sites/expedition/reproducing-our-ancestors/.

Riel-Salvatore, Julien. "A Spatial Analysis of the Late Mousterian Levels of Riparo Bombrini (Balzi Rossi, Italy)." *Canadian Journal of Archaeology* 37, no. 1 (2013): 70–92.

———. Interview by Lydia Pyne, phone, September 24, 2014.

"Rising Star Empire Cave 2014 Annual Report." SAHRA. www.sahra.org.za/sahris/heritage-reports/rising-star-empire-cave-2014-annual-report.

Rosny, J. H. *Quest for Fire*. New York: Ballantine, 1982.

"Rude Palaeoanthropology." *Nature* 442, no. 7106 (August 31, 2006): 957. doi:10.1038/442957b.

Sanders, Carol, ed. *The Cambridge Companion to Saussure*. New York: Cambridge University Press, 2004.

Sawyer, Robert J. *Hominids*. New York: Tor, 2003.

Schaaffhausen, H. "Teilen des menschlichen Skelettes im Neanderthal bei Hochtal." *Verhandlungen des Naturhistorischen Vereins der preussischen Rheinlande und Westphalens* 14 (1856): 38–42, 50–52.

Schindler, Karolyn. "Piltdown's Victims: Arthur Smith Woodward." *Evolve* 11 (2012): 32–37.

Shakespeare, William. *The Tempest.*

Shapiro, Harry L. *Peking Man: The Discovery, Disappearance and Mystery of a Priceless Scientific Treasure.* New York: Simon & Schuster, 1974.

Shen, Grace Yen. *Unearthing the Nation: Modern Geology and Nationalism in Republican China.* Chicago: University of Chicago Press, 2013.

Smith, Grafton Elliot. *The Evolution of Man: Essays.* London: Oxford University Press, 1927.

Smith, Pamela Jane. "Professor Dorothy A. E. Garrod: 'Small, Dark, and Alive!'" *Bulletin of the History of Archaeology* 7, no. 1 (May 20, 1997). doi:10.5334/bha.07102.

Sommer, Marianne. *Bones and Ochre: The Curious Afterlife of the Red Lady of Paviland.* Cambridge, MA: Harvard University Press, 2007.

———. "Mirror, Mirror on the Wall: Neanderthal as Image and 'Distortion' in Early 20th-Century French Science and Press." *Social Studies of Science* 36, no. 2 (April 1, 2006): 207–40. doi:10.1177/0306312706054527.

Spencer, Frank. *The Piltdown Papers, 1908–1955: The Correspondence and Other Documents Relating to the Piltdown Forgery.* New York: Natural History Museum Publications and Oxford University Press, 1990.

Spoor, Fred. "Palaeoanthropology: Malapa and the Genus *Homo.*" *Nature* 478, no. 7367 (October 6, 2011): 44–45. doi:10.1038/478044a.

Straus, William L., Jr., and A. J. E. Cave. "Pathology and the Posture of Neanderthal Man." *Quarterly Review of Biology* 32, no. 4 (December 1, 1957): 348–63.

Tappen, N. C. "The Dentition of the 'Old Man' of La Chapelle-aux-Saints and Inferences Concerning Neandertal Behavior." *American Journal of Physical Anthropology* 67, no. 1 (May 1, 1985): 43–50. doi:10.1002/ajpa.1330670106.

Tattersall, Ian. *The Last Neanderthal: The Rise, Success, and Mysterious Extinction of Our Closest Human Relatives,* revised ed. New York: Basic Books, 1999.

Than, Ker. "Surprise Human-Ancestor Find—Key Fossils Hidden in Lab Rock." *National Geographic News,* July 14, 2012. http://news.nationalgeographic.com/news/2012/07/120712-human-ancestor-fossils-sediba-science-berger-live.

Tocheri, Matthew W., et al. "The Primitive Wrist of *Homo Floresiensis* and Its Implications for Hominin Evolution." *Science* 317, no. 5845 (September 21, 2007): 1743–45. doi:10.1126/science.1147143.

Van Tuerenhout, Dirk. Interview by Lydia Pyne, in person, November 15, 2012, and May 12, 2015.

Waara, Anneli. "Unique Tooth Reveals Details of the Peking Man's Life." Uppsala University. www.uu.se/en/news/news-document/?id=4266&area=2,5,10,16&typ=artikel&lang=en.

Weiner, Joseph Sidney, Kenneth Page Oakley, and Wilfrid Edward Le Gros Clark. *The Solution of the Piltdown Problem*. London: British Museum (Natural History), 1953.

Williams, Lachlan. "Academia Is 'Bitchy': Fight Erupts over 'Hobbit' Fossil." 9 Stories, NineMSN, September 23, 2014. http://minisites.ninemsn.com.au/9stories/8909984/academia-is-bitchy-fight-erupts-over-hobbit-fossil.

"Wits Scientists Reveal New Species of Hominid." University of the Witwatersrand, April 8, 2010. http://kim.wits.ac.za/index.php?module=news&action=viewstory&id=gen11SrvoNme53_81569_1270732348.

Wong, Kate. "Could a Renewed Push for Access to Fossil Data Finally Topple Paleoanthropology's Culture of Secrecy?" *Scientific American*, May 8, 2012.

———. "Is *Australopithecus Sediba* the Most Important Human Ancestor Discovery Ever?" *Scientific American*, April 24, 2013. http://blogs.scientificamerican.com/observations/is-australopithecus-sediba-the-most-important-human-ancestor-discovery-ever/.

Woodward, Arthur Smith. *The Earliest Englishman*. London: Watts, 1948.

Yardley, William. "They Didn't Love Lucy." *New York Times*, March 13, 2009. www.nytimes.com/2009/03/19/arts/artsspecial/19bust.html?_r=0.

Yen, Hsiao-pei. "Constructing the Chinese: Paleoanthropology and Anthropology in the Chinese Frontier, 1920–1950." Doctoral dissertation, Harvard University, 2012. http://dash.harvard.edu/bitstream/handle/1/10086027/Yen_gsas.harvard_0084L_10240.pdf?sequence=1.

Zipfel, Bernhard. Interview by Lydia Pyne, in person, July 1, 2013.

INDEX

Page numbers in *italics* refer to illustrations.